Praise for *In the Company of Bears*

"Ben Kilham's *In the Company of Bears* is surely the most insightful book about animals written in the last one hundred years. His observation of black bears is the best ever done, his data are flawless, and these attributes have created a landmark of science that as far as I know has not been equaled with any other species. And if that's not enough, it's also a page-turner and a must-read. It left me breathless."

—Elizabeth Marshall Thomas, author, *The Hidden Life of Dogs* and *The Tribe of Tiger*

"Like Jane Goodall's studies of chimps, Ben Kilham's work with black bears is more than just revealing: it's revolutionary. This riveting book supports two astonishing conclusions: that bears are far more sophisticated than most scientists dared imagine, and that dyslexia, once considered a failing, may simply be another, and often valuable, way of thinking. Ben's work will transform our understanding of how animals live—and how science should be done."

—Sy Montgomery, author, *Walking with the Great Apes* and *Search for the Golden Moon Bear*

"*In the Company of Bears* is a brilliant revelation about black bears and a paean to human potential. After decades spent caring for orphan bears and releasing them into the wild, Ben Kilham, a dyslexic animal lover, has now summarized what he has learned about their rich social complexity and humanlike intentionality. The result is to turn a supposedly familiar species into a creature of unsuspected acuity. Part science, part intuition, this enticing natural history is a provocative argument about animal minds, and an intimate celebration of life in the New Hampshire woods."

—Richard Wrangham, author, *Catching Fire: How Cooking Made Us Human*; codirector, Kibale Chimpanzee Project

"Magnificent! *In the Company of Bears* is a brilliant read. Kilham perfectly exemplifies how much the world has to gain from the exceptional insights of dyslexic individuals, who often possess a special talent for finding order hidden in the complex patterns of the real world. We cannot recommend this book highly enough."

—BROCK AND FERNETTE EIDE, authors, *The Dyslexic Advantage*

"No one knows black bears like Ben Kilham does. During the past decade alone he has raised and rehabilitated to the wild more than one hundred orphaned cubs. The social code of the bears, as so delightfully described in *In the Company of Bears*, includes cooperation, imitation, fairness, punishment for infraction, reciprocity, and other traits of this solitary yet highly social species. Human parallels are drawn, too. The charm and core of the book lie in its anecdotes and unique insights, especially of the female Squirty, with whom Ben has had a friendship for more than seventeen years. Read this fascinating book and see the bear's world with new eyes."

—GEORGE SCHALLER, author, *The Last Panda*;
science director, Wildlife Conservation Society

"*In the Company of Bears* by Benjamin Kilham is one of the finest books on the natural history of an animal that I have ever read. As gripping as the best novel, it is very readable and provides great insights into the mind of the black bear and the human. The research that Kilham has done on the behavior of black bears is groundbreaking and will lead us to a much better understanding of the development of intelligence in mammals and the evolution of cognition throughout the animal world."

—JAMES R. SPOTILA, Betz Chair Professor of Environmental Science,
Drexel University; author, *Saving Sea Turtles*

"Kilham's latest is the most thought-provoking book that I've read about bears. It not only made me think differently about black bears, it also enriched how I feel about them."

—STEPHEN HERRERO, biologist and author,
Bear Attacks: Their Causes and Avoidance

In the Company of Bears

What Black Bears Have Taught Me about Intelligence and Intuition

BENJAMIN KILHAM

Chelsea Green Publishing
White River Junction, Vermont

Originally published in 2013 as *Out on a Limb.*

Editor: Joni Praded
Project Manager: Bill Bokermann
Copy Editor: Laura Jorstad
Proofreader: Helen Walden
Indexer: Barbara Mortenson
Designer: Melissa Jacobson

Printed in the United States of America.
First printing October, 2013
10 9 8 7 6 5 4 18 19 20 21

The Library of Congress has catalogued the previous edition as:
Kilham, Benjamin.
Out on a limb : what black bears taught me about intelligence and intuition / Benjamin Kilham.
pages cm
Includes index.
ISBN 978-1-60358-390-9 (hardcover) – ISBN 978-1-60358-391-6 (ebook)
1. Black bear–Behavior–Anecdotes. I. Title.

QL737.C27K568 2013
599.78'5–dc23

2013025262

Chelsea Green Publishing
85 North Main Street, Suite 120
White River Junction, VT 05001
(802) 295-6300
www.chelseagreen.com

To my loving wife, Debbie

CONTENTS

FOREWORD

Benjamin Kilham has hand raised over a dozen orphaned bear cubs and then reintroduced them into the wild by walking with them in the forest, giving them the opportunity to gradually learn how to find food and identify the smells of other bears. Eventually, he released these cubs into the wild, where they had complex social interactions with other bears. When they had their own cubs, they still viewed Ben as their mother, and he would go into the woods and visit them as a member of the bear society. In the process, he learned amazing things about both the orphans he reared and other wild bears in his study area. And as a result, this fascinating book has detailed descriptions of bear body language, oral communication, and behavior–and how Ben learned to read them.

I can relate to Ben and his story because his dyslexia and my autism have made us both visual thinkers who are very observant of small details that most other people miss. Animals live in a sensory-based world, and if you want to understand them, you must get away from the confines of verbal language. In my own work, designing livestock-handling facilities to improve animal welfare, I noticed that cattle are extremely sensitive to visual details in their environment. During my very first observations of cattle handling, I noticed that the animals would stop and refuse to walk over shadows, or balk at reflections on shiny parked cars or in a mud puddle. Similarly, Ben learned how to read ear position on bears to determine how they were feeling. Yet visual details that are obvious to Ben and me are often overlooked by many people. In my livestock behavior class, for instance, I show my students many pictures of visual distractions that cattle will notice—such as a chain hanging in a chute or a coat on a fence—and explain that if the distractions are removed, the cattle will move easily. The following week, we have a lab where we handle cattle in a chute that is used for veterinary work. Since part of my goal is to teach my students to be better observers, I test their powers of observation by deliberately leaving a swinging chain in the chute. When I ask the students if they saw the swinging chain, about 80 percent had failed to notice it, even though a picture of a hanging chain had been shown the previous week in class.

When I first started my cattle work in the 1970s, I did not realize that my mind worked differently. I assumed that everybody thought in pictures. Gradually, as I asked other people how they thought about different things, I learned that my thought processes were different. When I asked people how they accessed their memory of common objects such as church steeples, for instance, I was surprised to learn that many recalled a generalized image and I saw only images of specific steeples. In other words, my concept of a steeple is based on many different specific steeple images put in the "steeple" file folder in my brain. I also discovered that there are degrees of visual thinking: for those who recall generalized images, the level of detail in those images varies from person to person. (You may wonder why I asked about steeples. The reason is that most people will vividly see their own house or car, but I discovered the differences in thinking styles when I asked about an object that everybody sees but they do not own.)

Both Ben and I are also bottom-up thinkers. When bottom-up thinking is used, small details are put together to form a theory. This is the opposite of the more typical top-down thinking. In top-down thinking, a theory is formed first and then the specific data are forced to fit it. To the contrary, Ben's methods for observing bears are similar to the groundbreaking research styles of Jane Goodall and Niko Tinbergen—as well as Konrad Lorenz, who raised baby geese and observed their behavior.

As Ben notes in the pages ahead, his dyslexia initially prevented him from getting a higher degree, but his ability to read emotion, notice minute details, and think in pictures rather than words have been key assets for studying and understanding animal behavior. Things that seem obvious to him are not obvious to other people—and that is partly due to the fact that he does not overgeneralize, a top-down-thinking mistake that many people make when trying to understand why a child or an animal behaves a certain way.

Over and over at my lectures, I get vague questions like "What can I do to solve my child's behavior problems in the classroom?" Or, "My horse goes berserk; how do I fix him?" To answer these questions, I need lots more information, including a detailed description of the actual behavior and what events or environments triggered it. After asking about five more questions about the horse, I learned that he went berserk in only one specific place—the cross ties where he would be held in place to be saddled. He was fine in all other situations. So, I suggested that he probably had had an accident in the cross ties and was now afraid of them. Since cross ties are essentially two ropes that keep a horse stationary between two posts, I suggested tying him up with just one lead rope in a new place. Animals will often get place-specific memories. They may also fear something they were seeing or hearing the moment something bad happened—and the fear memories are often stored in memory like a picture. For example, there is a dog that is afraid of men wearing baseball hats, but if a man wearing a baseball hat takes it off, the dog will accept him. That doesn't mean the dog is afraid of the hat, though; it more likely means the dog associates the image of a man wearing a baseball hat with a troubling event. Such is the way of thinking in pictures.

For years, Ben has been an academic outsider because he observed animals to understand how they behaved, and why, rather than by performing experiments. Yet, observation is an important part of science, one that we accept readily in fields like astronomy. Imagine what would happen if we demanded a control for all astronomical observations? The control for findings from the Hubble Space Telescope might be to point the telescope at the ground. That would be silly.

Some animal-behavior specialists also criticized Ben for hand raising cubs to gradually introduce them to the wild. Finally, though, he was invited to become an academic insider by publishing his work as a PhD thesis. The academics had a change of heart after Chinese researchers started using Ben's findings and methods to form their own plans for reintroducing pandas into the wild.

Different minds work in different ways, and we need to find ways to foster a variety of talents. I am very concerned that many talented visual thinkers are going to be increasingly marginalized in today's educational system. Ben forged ahead and did what made sense to him, despite tough times in the academic world. As a result, he has unveiled their wild world for us, helped orphaned bears reenter it, and helped solve human–bear conflicts.

–Temple Grandin
Professor of Animal Science, Colorado State University

ACKNOWLEDGMENTS

I could not have done any of this work without the help of a lot of other people. I would like to thank my wife, Debbie, for understanding, financial support, and help when needed with the cubs. My sister, Phoebe, who has been a partner and primary caregiver with the cub rehabilitation from the beginning. My brother, Josh, who has helped whenever needed, and my sisters-in-laws, Susan Kilham and Sara Chaffee. Andy Timmins, Will Staats, Mark Ellingwood, Kris Rines, Steve Weber, and the other members of the New Hampshire Fish and Game Department, and Forrest Hammond of the Vermont Fish and Wildlife Department, without whose cooperation none of this would be possible. Don Cooke for introducing me to GPS and Lee Larson for helping with Squirty Cam. Pete Pekins, Adrienne Kovach, Stephanie Koster, and Wes Smith from the University of New Hampshire for their support and help with genetics and GPS. Jack Hoopes from the Dartmouth-Hitchcock Medical Center for help with anatomy and pathology. Interns Peter Abdu and Jonathan Trudeau for their help and dedication. Peter Tse and Walter Sinnott-Armstrong for our discussions. Richard Wrangham and Richard Estes, who gave advice on behavior and anatomy. George Schaller for his support over the years and for agreeing to be on my graduate committee. Steve Herrero for his help and support. Sy Montgomery for all her assistance over the years.

Jennifer Unter from the Unter Agency for finding the right publisher. Hillary Rosner for help with my book proposal. My editor, Joni Praded, for pulling the book together, and everybody at Chelsea Green Publishing. Bill Greene and Lauren Gesswein for the use of their photographs. Roy Day for helping with projects. Tony and Sue Ryan for access to my woodlots through their property. Andy Lumley and Tim Chow for flying to look for lost bears. Bill Wanner and Jim Dickerson for helping when needed. And to all of those whose contributions I cannot think of at the moment.

Special thanks to James Spotila and Drexel University for my acceptance into the doctorate program. To John Spotila and Sheri Yi of the Global Cause foundation for my trips to China and introductions to the giant panda. To Bob and Sandy Green of Green Woodlands Foundation and David and Barbara Roby of Bear Hill Conservancy Trust for their friendship and support. To James Jukosky and to Susan Jukosky and their families and to the Charles E. and Edna T. Brundage Charitable, Scientific, and Wildlife Conservation Foundation. And to all of those who have provided generous support for the bears.

Finally, thanks to the landowners whose forests I've roamed tracking bears, and to all of those who are learning to live with bears.

INTRODUCTION

Late on a spring afternoon, at the usual time, I drove my truck along a dirt road a mile deep in the New Hampshire woods, up the steep rocky inclines of Smarts Mountain, slick with mud from the lingering spring runoff. After half an hour, I reached my destination, a small clearing at the site of a farmstead abandoned in the 1860s. A young hardwood forest, full of red oak, white birch, and sugar maple, had regenerated on a former pasture amid a scattering of large white pines. I have come to call this place the Lambert clearing. The old farmhouse's original lilac bush was preparing to bloom—as were wild apples, black cherries, blueberries, and many more plants that would provide summer food for bears. Red oak flowers, which would yield the next year's nuts, dangled around tiny acorns destined to drop in the fall.

I parked my truck and took a few steps into the clearing, softly calling to announce my arrival in case the breeze was blowing my scent in the wrong direction. Fresh moose tracks scarred small patches of clover. A black-and-white warbler picked at tiny insects among the apple blossoms, and a flock of chickadees darted by. Two yellow-bellied sapsuckers were loudly resolving a territorial dispute. After a few minutes, something black and moving caught my eye. It was a fourteen-year-old, 180-pound black bear. Trailing behind her was a small male cub, rocketing along to keep up with his mother's lumbering gait.

The bear stopped at the edge of the woods, raised her head to the breeze, and opened and closed her mouth. She was monitoring the clearing for other bears. Satisfied, she continued along the trail toward me, pausing to munch some tender sprouts of golden alexanders. I stood still, not wanting to spook the little cub. The mother bear approached me calmly, an expectant look on her face, eyes wide and ears forward. She came up right beside me, put her paws on my arm, and stood up on her hind legs. Measuring five feet from tip of nose to tip of tail, she stood with her head at my chest. I leaned over slightly, letting her sniff my mouth to inspect my breath. It was both a greeting and an investigation of what I had recently eaten, a behavior I've now seen thousands of times among bears, but which hadn't previously been recorded in bear science. When she was done, I gave her a single Oreo.

Before the bear even had a chance to put her front paws back on the ground, she caught a scent in the air. Her upper lips extended and pursed into an oval. Her eyes glared. This, for a bear, was a look of terror. She spun around to find her cub, but he had already sensed her concern. He raced for the nearest large pine and scampered into its highest branches. His mother stood at the base of the tree and waited.

A full ten minutes passed before the threat became evident–at least to me. A large breeding male entered the clearing. As soon as he arrived, the mother bear climbed the tree as well, stopping halfway between the ground and the cub. Most breeding males know enough to stay away from sows with young cubs, but this one had broken the rules. He approached the big pine but stopped beside a smaller tree nearby, rubbing his back against it while standing on his hind legs. His message was clear: "I am available."

She did not need to mull it over. Immediately she went into free fall, dropping forty feet to the forest floor. She hit the ground running in full assault mode, and chased the much larger male off into the forest. I watched the scene play out, standing next to my truck, knowing that I was in no danger because the female bear approved of my presence there. She'd let me know when she wanted me to leave.

Before you dismiss this as a scene out of *Grizzly Man*, let me set the record straight. I am, in essence, that bear's mother. Her name is Squirty, and I've known her since she was a seven-week-old cub, weighing just three pounds. She was separated from her real mother in 1996, when a logging operation disrupted their den. I raised her, along with her two siblings, and released them back to the wild when they were a year and a half old. Squirty and her brother and sister were among the first few sets of orphaned black bear cubs that I took under my wing as part of an arrangement with state wildlife agencies to try to give these animals a chance to survive in the wild. I chronicled my experiences with those early cubs in my first book, *Among the Bears*, published in 2002. Since then, more than one hundred cubs and juveniles orphaned by logging operations, car accidents, hunting, and other causes have ended up in my care. But this is only part of my work with bears.

For nearly twenty years now, I've been interacting with wild black bears as a state-licensed bear researcher. My work is safe, methodical, and officially sanctioned. Nevertheless, I'm not a scientist. Severe dyslexia caused me to stop my academic career after getting a bachelor of science degree. But while I have had to overcome my difficulty reading books, I have never had any trouble reading nature. And this ability has proved enormously useful in studying bears.

I've been able to observe these animals closely and continuously, and have amassed a wealth of new information about the astonishing ways that black bears communicate, socialize, and share resources. Yet as much as the bears have taught me about their own world, they've also helped me gain important insights into ours. You don't need to be a credentialed scientist to make new discoveries. You just need to go outside and open your eyes.

Bears, it turns out, are a lot like humans. They form alliances with strangers, they make calculations about relative costs and benefits, they lay down rules and punish those who break them. They trade based on a clear system of reciprocity. They communicate using equal parts emotion, intention, and dependence on context–a combination that is essential for communication between strangers and in fact forms the basis for language.

I learned about bear society using radio collars, remote cameras, and DNA testing. But the most dependable and enduring method I've employed has been observing the bears with my own eyes. This isn't always easy. Black bears are shy, secretive animals who live in dense woods, where observation comes with its own set of physical challenges. That's why it's taken me all these years to piece together a model of their world. In the summer, visibility in a New Hampshire forest is often less than seventy-five feet, meaning the chances of watching a bear are essentially nil. It's the reason most of the bear sightings you hear about take place in someone's backyard, when the animals come looking for a hearty meal from a bird feeder or garbage can. But by watching bear cubs grow up, following them to adulthood, observing one–Squirty–as she established her home range, and monitoring dozens of others who move in and out of that range, I've had a front-row seat to the lives and behavior of a remarkable species. I've been able to study many individual bears up close and continuously. It's taken a little luck and a lot of effort.

Year after year, I've gone into the woods and watched. It was all I could do, really; I had no funding and no university supporting me. I also had no reputation to worry about protecting, and no hypothesis that I needed to prove. I had a love of nature, a deep curiosity about animal behavior, and an ability to tap into animals' emotional sensitivity–which I perhaps developed in lieu of a capacity for learning from books. It is not that I don't read, because I do. I read everything I can find that relates to bear behavior specifically, and animal behavior in general. I can figure out how things work by observation alone, but in order to share what I've learned with others I have to understand the significance of my observations and apply the words of science. My perceived shortcomings turned out to be a perfect fit for studying bears; they enabled me to gain important insights into bear behavior, and also to see the value of alternative ways of learning and experiencing the world. Instead of cutting myself off from my senses and intuition, as the scientific method often encourages, I've relied on those capacities–using them to enhance the rational side of my brain. Too often we pit instinct and reason against each other, when in fact combining them can help us tremendously in our quest to understand the natural world.

I didn't set out to test any preconceived theories. I simply bottle-fed and raised the orphaned baby bears whom wildlife officials brought to me. I did the same for about eight orphaned juvenile bears who had no experience outside the den with their natural mother. My sister, Phoebe, and I walked these cubs in the forest to give them as much of an education as possible. When they were old enough, I worked with wildlife officials to return the bears to the forest and then continued to monitor them. I watched "my" bears, I watched their offspring and mates, I watched the other bears with whom they share their landscapes. As a result, I've been able to study bears in fenced enclosures on my property while they are being rehabilitated, and I've observed some of the bears that I returned to the wild on several hundred acres I own on the outskirts of my town. (Others have been released elsewhere in the forests of New Hampshire and Vermont.) My life revolves around this patch of land, where I not only observe bears but also operate a tree farm and tap trees for maple syrup.

Each year, from late April to November–give or take a few weeks, depending on how the weather influences the bears' hibernation schedule–I meet up with Squirty most evenings in the clearing. We're often joined by other bears, who live and feed in the area at Squirty's discretion and abide by the rules she lays down. Some of these bears are Squirty's female descendants, her daughters and granddaughters. Others are unrelated.

When I started my bear project, I had no great expectations. I was intrigued by the carnivores that lived in my local woods–not just bears but otters, bobcats, and fishers, too. I hoped one day state officials would need my help in caring for one, but until the first cubs arrived I had mainly been called upon to rehabilitate crows, ravens, and other small birds and mammals. Still, my interest in these top-of-the-food-chain animals was just a personal fascination. I certainly didn't expect to break any new ground. Bears, after all, were high-profile, charismatic animals that, by all appearances, had been studied extensively. There were dozens of books and hundreds of journal articles about them. Surely everything there was to know had already been discovered. But my father–a doctor, researcher, and naturalist–gave me some good advice when I was a boy. "Discovering something new is great," he said. "But the real reward lies in discovering something for yourself." I wanted to understand these animals' behavior for myself, through firsthand experience. As it turned out, though, I discovered a great deal of new information–about black bears, about myself, and about human behavior.

The conventional wisdom has always been that bears are solitary creatures. But my research shows that in fact they are highly social animals with complex relationships and a well-developed system of justice. On many occasions, I've been a participant–however unwittingly–in the bears' judicial matters. Every aspect of black bears' behavior is rooted in the need to share food in landscapes where it is unevenly distributed, available at times in huge surpluses and at other times in potentially devastating shortages. This is similar to the world in which humans are thought to have first evolved, before we learned to create our own surpluses.

Some of my findings about bears contradict the standard thinking–about black bears and about animal communication in general. I don't use conventional methods; my research is primarily qualitative, much like Jane Goodall's was, based on the accumulation of hundreds upon hundreds of diverse observations rather than standard quantitative science. Because of this, I have difficulty publishing my results in scientific journals. Still, I'm invited to lecture at universities and present at conferences, I'm beginning to work with graduate students, and I've even been called upon to help the Chinese reintroduce pandas into the wild based on my experience reconnecting black bears with their natural environment. But my outsider status has kept my research largely on the fringes, even as my well-documented evidence has piled up.

So, I have written this book to share what I've learned over the years about black bears and their society, and also about what these large-brained omnivores can teach us about ourselves. Despite all our technological advances, most of our actions as humans can still be described and predicted by the simple terms of animal behavior.

But I have other reasons, too, for picking up the pen to write about bears once again. The first is to avoid unnecessary conflict. Now, more than ever before, people are encroaching on bear habitat, pushing the boundaries of urban areas farther into the forest, and that means that human–bear interactions are on the rise. An understanding of these powerful mammals' behavior is essential for minimizing conflicts and ensuring their survival. The second reason is personal. My years with black bears have taught me a tremendous amount about myself and about the importance of trusting in my natural skills rather than dwelling on my shortcomings. Through observing bears, I have learned to turn my dyslexia into an asset instead of a liability.

For Squirty and her fellow bears, information can only come firsthand. Observation is the only method for amassing new knowledge; images become a means of storing information. If a bear makes the wrong choices, it may be punished–and the ultimate punishment could be death. Making the right decisions is vital, so bears pay close attention to the information available to them. I've practiced this same behavior in my bear research, and it has led me to some surprising discoveries.

In the bears' world, as in ours, hierarchies of dominance develop–based on age, family status, physical size, and personality, as well as over food, security, and the right to mate. These issues are dealt with on an individual basis, and maintained with various forms of communication. Being included in some of these activities has allowed me privileged insight.

I hope to share, in *In the Company of Bears*, what I've learned about understanding and recognizing systems and patterns, in both nature and ourselves. Everything in nature happens for a reason, and every mechanism has a purpose. The world of black bears–their social system–turns out to be more complicated than people assumed. So how does it work? Where did it come from? Why do these animals do the things they do?

Come take a walk in the woods with me and let's find out.

CHAPTER ONE

Walking with Cubs

Winter came late in 2012, and by the time sporadic frosts turned to steady cold and the calendar flipped to 2013, I found myself with a unique problem. For nearly a year, my life had been ruled by a growing cadre of orphaned black bear cubs. That alone was not an issue. Over the last twenty years, I have been raising orphaned cubs whom state wildlife officials bring to me when their mothers have unfortunate run-ins with cars or guns, or leave them behind when food is scarce or dens are disturbed. I have learned to bottle-feed cubs, wean them onto solid food, and clean up after them. I've managed their transitions from their earliest days in my home to their eventual move to an outdoor enclosure and their final move back into the wild. Caring for bears has even become a family affair, involving my wife, Debbie, and sister, Phoebe. But we had been used to, at most, looking after two or three cubs at a time. So it was a surprise when we found ourselves with twenty-seven young and hungry bears. It was an even bigger surprise when winter set in and twenty of them refused to do what bears do and take a long winter's nap. Like kids at a slumber party, they were too busy rollicking around with one another to let biology take over and lull them into deep sleep.

This was new: black bears who would not hibernate.

And so Phoebe and I found ourselves playing mother to an unruly but happy pack of juvenile ursines who feasted on apples, dried corn, acorns, and dog kibble while hanging out together in an eight-acre fenced and wooded enclosure on my property in Lyme, New Hampshire–a tiny outpost in a region of the state that hugs the Connecticut River and spans out into small towns and large tracts of forests that hundreds of wild black bears call home. We tried holding back on food to no avail. Snowstorms, ice storms, and even a nor'easter failed to entice them to settle down. At night, after climbing in trees and playing with one another on the forest floor, the bears would retreat to various dens in brush piles and structures scattered throughout the eight-acre fenced patch. In a normal year, they'd just sleep in the den throughout the winter.

Our bumper crop of bears was rooted in an unfortunate twist of natural events. Two years before, the New Hampshire forests were flush with beechnuts, wild apples, raspberries, blackberries, blueberries, winterberries, chokecherries, mountain ash, and other berries galore–all foods bears love to eat. This banner food year translated into a banner mating year, and sows feasting on the abundant wild crop produced an equally abundant number of cubs. But for many bears, the happy story did not continue. An early warm spell the following spring caused the wild apple crop to fail. There was a shortage of berries, too. Over the coming months, there were far more bear mouths to feed than the forest could handle. Some sows were forced to abandon cubs. Others had to go farther afield to find food. Sometimes that meant raiding bird feeders or chicken coops, scenarios that often ended poorly for mother bears and their then-orphaned cubs.

By the time mid-March rolled around that spring, we already had three newborn cubs in our care. We had taken in many cubs of various ages over the years, but it had been some time since Debbie and I shared our home with newborns. I had long ago learned what being a bear mother–at least a human one–was all about. It was at least an eighteen-month affair that began with setting up a nursery in our spare bedroom, bottle-feeding the cubs around the clock, letting them suckle on earlobes and fingers despite ever-growing teeth and sharp claws, and giving them a mix of free range and protection as they moved on to the rambunctious stage that can only be equated with human toddlerhood. Eventually, when they grew too feisty for their pen in the house, we would move them into the barn I'd built as a bear rehab station, where Phoebe, Debbie, and I would feed them daily. Often, when they were old enough to begin learning the ways of the woods, I would accompany them on daily walks through the forest, giving them a chance to learn in their natural environment. Sometimes Phoebe would do the same. We'd watch as they'd fall into the natural routine of rooting in the soil for acorns, or master their first tree climbs.

It had been nearly twenty years since I had received my first bear cubs, Little Boy and Little Girl (or LB and LG, as I came to call them) and begun the journey of figuring out just how to raise and reintroduce wild bears. I had suffered my first parental pangs with LB and LG, deep joy when they allowed me to experience their world on our daily walks, and deep heartache, too, when LG (after unwanted visits to a neighbor) suffered a fatal wound from a plastic bear repellent and LB was claimed by a hunter. Other infant bears had later come into our care, and we repeated the cycle. But by the time the wooden pen in our spare bedroom was filled with three newborns in that unusual spring of 2012, we knew we were in for one of our biggest adventures yet.

We quickly got reacquainted with round-the-clock bottle feeding. Among the crew were Clarkie, who was an easy feeder, and Big Girl, who was a stubborn eater at first but eventually learned to grasp the rubber nipple of the bottle and nurse, gulping down the lamb's-milk-and-yogurt mixture that we had perfected over the years to replace their mother's milk. Like human children, they bawled when they were hungry and were sweet and content when satisfied. It wasn't long before the cubs began to play with one another. They wobbled dramatically as they tried to stand on all fours—a trait often mistaken for an ailment but one that's actually protective, preventing them from wandering away from a winter den.

Soon they were false-charging their shadows inside their wooden enclosure. At feeding time, we'd let them loose on the floor and watch them as they began to entertain vertical movement and show an interest in climbing. They began to climb up my chair legs and pant legs (something that taught me, long ago, the value of wearing two layers of heavy pants around cubs). When all were running and climbing, that signaled the move to the barn. By the time eastern flycatchers and yellow-bellied sapsuckers returned to the surrounding woods, the cubs were fully capable of havoc. They ran, wrestled, and played until exhausted–stepping up the action when Phoebe, Debbie, or I was in the barn, as if our presence provided extra security for them to test their boundaries. As the weeks wore on, they exhibited signs of maturing. They'd flash a quick temper or a bite when asked to do something against their will, not unlike a human child who throws a tantrum to get his or her own way. The simple act of removing them from a lap they wanted to climb into could result in a roar and a quick nip.

When they were all comfortable in their pens, downing full bottles in the morning, suckling one another's ears, and generally carrying on, the time was right to begin taking them for short walks. It was a hot, dry spring, and we first struck out on an eighty-five-degree day in April. As we ambled along, the cubs mouthed coltsfoot flowers and striped maple buds; they smelled my breath when I chewed on beech buds and poplar leaf starts. Other orphaned bears, all yearlings who had been in our care over the winter, were out in the forested enclosure eating willow buds and nodding sedge. But as the spring wore on, even more cubs arrived.

Students from Tufts Wildlife Clinic soon sent us Monty and Slothie, who, despite being the same age as the other cubs, weighed just two and three pounds, respectively. It's not unusual for first-time mothers to run out of milk due to lack of fat and abandon their young, and that's likely what happened to Monty and Slothie's mother. In normal years, a small, underfed cub would succumb to the cold, but in this warm spring Monty and others like her were able to make it to people's yards and bird feeders, where they were discovered and brought in. Another cub,

Little E, arrived equally small, hypothermic, dehydrated, and covered with ticks. When they joined the group, they ended up at the bottom of the heap in most of the bear pileups, but soon were greeted with lip-to-lip kisses and soft chirps by some of the others. It was not long before they, like the others, were exploring the forest, climbing trees, swimming in the pond, and occasionally falling off rotted logs and retreating to the comfort of my shoulders before daring to venture out again. Like other abandoned cubs, they showed gratitude. They would come right up to my face, look me in the eyes, and give me a kiss on the lips to ensure I would care for them.

By the time I was putting round barrels baited with food in the forested enclosure so that I could start trapping the yearlings for release, news of even more cubs arrived. Then Debbie and I got a call that another three cubs had lost their mother at a chicken coop, and two more had been orphaned by a car accident. Three more arrived after their mother was shot at a beehive in Vermont, and then six more came in from additional deadly chicken coop encounters in the far reaches of New Hampshire. In the fall, eight more cubs showed up at various locations throughout the state and so joined our crew in Lyme. One, of these, a seven-pound cub, had just shown up at a Dairy Queen in the White Mountains.

Surely, we had never seen a year like this one, but alongside the day-to-day care of the cubs, I also needed to get on with my other bear business, which involved releasing the yearling bears into suitable habitat, and studying the seventeen wild bears that I'd been tracking and observing for years in the hope of learning more about their lives.

Ever since raising Little Boy and Little Girl, I was increasingly drawn into the bears' world. From the beginning, questions emerged about how bears behaved and why–questions whose answers came slowly, as the information and evidence began to pile up and patterns emerged. Though I certainly didn't know it at the start, the cubs' behavior would become powerful backup evidence for much of what I observed later among adult bears. Yet I had to spend two decades learning how bears communicate in order to finally make sense of the whole system.

I also had to grow comfortable with conducting my own personal experiment–doing things my way with no expectations of success, only to learn the way I was born to. In preparation for raising my first set of cubs, I searched the scientific literature to see what had been written on the subject. There were only two studies related to rehabilitation or behavioral research with black bear cubs. One was by a scientist who released six-month-old cubs to the wild and found that they were still alive three months later. He concluded that orphan bear cubs older than five months could survive by themselves in the wild. The study had a major impact in the wildlife community and was used to justify leaving orphan cubs in the wild or returning rehabilitated cubs to the wild at that very young age. I was stunned. Even as an amateur, it was clear to me that the study was simply drawing a convenient conclusion.

Wild bears spend eighteen months with their mothers. Surely there must be a reason for this. How could anyone come along and just say, "Hey, let's take away two-thirds of that time and I bet things will work out fine"? It's counterintuitive. The bears' lengthy cub-rearing strategy is the result of more than five hundred million years of evolution–and yet scientists believed that it could be reduced to five months? The biggest stress for any bear is surviving the winter, but the study did nothing to explain how the cubs were going to build up the required fat reserves without their mothers to show them how to find food or fend off competing bears. This was perhaps my first real exposure to the fact that, in many ways, we have become so detached from the natural world that we can't even trust our own instincts. We look for meaning in other people's theories, even when those theories directly contradict what we can see with our own eyes.

The nice thing about studying natural systems, though, is that when in doubt, you can always go back to nature: I was determined to find out what cubs learn during those eighteen months. I discovered distinct developmental stages that clearly affect bears' ability to explore–and survive–on their own. I also formed a relationship with one of these cubs, Squirty, that would last well into her adulthood–offering me a rare window into her world and the world of other wild bears around her.

One spring day back when Squirty was a cub, I climbed a steep trail with her and her two siblings, Curls and The Boy. The bears were seventeen months old, almost ready to leave home, and I knew from experience with the first set of cubs–Little Boy and Little Girl–that it was time for them to interact with wild bears, even if it meant they'd spend the night out in the woods instead of safe in their fenced enclosure on my property. I led the way up the trail, confident that the cubs would later retrace the route without me. The next morning, off they went by themselves. When I went to look for them, using the signal from The Boy's radio collar to track them down, they raced off down the other side of the mountain as soon as I got within earshot. That night, the cubs did not come home.

The following day was foggy and rainy, and the wind kept my scent away from the cubs. I snuck back up the mountain, and there they were, feeding in a beech stand. When they finally heard me–I stepped on a twig, which broke with a loud snap–they ran up the nearest tree, terrified. That's when I saw their newfound friend, an adult female wild bear who was not about to let me harm the cubs. She came at me in a false charge, a common behavior that says, unequivocally, "Back off." This wild bear was protecting my own cubs from me!

Just then the wind shifted, and the cubs caught my scent. Down they scampered, making the noise I've come to know as the "moan of recognition." The wild bear simply sat down and watched. The cubs rubbed all over me, masking my scent to prevent me from scaring off their wild friends. (Amazingly, afterward I was able to walk around that beech stand with wild bears paying me no attention.) And then each one in turn bit me on the arm, hard enough to leave a bruise but not to break the skin. They were punishing me for interfering with their friends–in much the same way your daughter or son would be furious if you decided to hang around during their teenage party.

This was just one of the countless remarkable events that took place right from the day the bears first came into my life. They were events that reshaped my life as much as they reshaped my understanding of and confidence in myself, something that had been shaky since my earliest school days.

In fact, my trouble with school began back in kindergarten. Teachers labeled me "disruptive," among other problems, and refused to let me continue to first grade. From then on, I struggled in both public and private schools, earning C's and being told year after year that I was bright but not applying myself.

I made it through college, earning a degree in wildlife biology from the University of New Hampshire. But I was unable to fulfill my dream of studying animal behavior in a graduate program. I had made up my mind about what I wanted to do in life, yet it was inaccessible to me. It wasn't my ability or understanding of biology that kept me out. I scored 720 of 800 on the biology test of the Graduate Record Exam. But my score in English was 400.

So I followed my other interest: gunsmithing. I headed west and enrolled in the Colorado School of Trades. There, I flourished. Everything was suddenly easy. I excelled at machine work, woodworking, welding, mechanics, and problem solving. My work as a gunsmith, gun designer, and inventor was rewarding in a way I had never before experienced. Nobody told me how to do things; I just did them. I could mentally conjure and manipulate designs, and then I could physically create them out of metal or wood. It was satisfying and tangible.

I discovered that my mind gravitated toward systems–not theoretical systems in books but real-life systems: the inner workings of guns, machines, social relationships, nature. Not only was I attracted to them, but I could understand them, remember them, and build on them. It was all about solving problems. I wanted to know how things work, how the parts fit together into the whole.

Temple Grandin, the autistic animal behaviorist, has described a similar approach to learning. In her book *Animals in Translation*, she writes about how, as a child, she tried to understand what made a dog a dog. "All the dogs I knew were pretty big," she writes, "and I used to sort them by size." So when her neighbors brought home a dachshund, she was befuddled. "I kept saying, 'How can it be a dog?'" she writes. "I studied and studied that dachshund, trying to figure it out. Finally I realized that the dachshund had the same kind of nose my golden retriever did, and I got it. Dogs have dog noses."

My bear research proceeded along similar lines. I wanted to understand how bears communicate, so I spent countless hours watching them rub their backs against certain types of pine trees until eventually it became crystal clear that they were signing the guest book or etching their name in the school desk: "Squirty was here." I learned, over years of collecting information, that bears respond in highly specific ways to different scents. Sometimes they might smile, other times they'll immediately make a false charge; they can quickly tell who left the scent, as well as that bear's social status.

I was forty years old–married, running a gunsmithing shop and a maple syrup orchard, and rehabilitating wildlife on the side–when I finally acquired a name to attach to my problem: dyslexia. Testing confirmed that I had a rocket scientist's IQ but a third grader's reading skills. I could read and comprehend, but I couldn't do it quickly. It might take me five times as long as a normal reader to get through a book. Words are images to me, and I often find myself projecting the wrong word in situations of similar-looking words like *beast* and *breast*. With meaning way out of context when words are confused, I have to go back and reconstruct the sentence.

Yet there seems to be a kind of equilibrium to the human mind, in which deficits are balanced with assets. We all have our strengths and weaknesses. People who are not good readers often are better at "reading" real-world situations. My father served in General Patton's army in World War II. Patton, it turns out, was dyslexic. He was at the bottom of his class at West Point. Yet when he got on the battlefield, he just blew across Europe. He could understand what was going on and intuitively act on it.

In my case, I have always been able to read and relate to emotional signaling, in both humans and animals. My father was always taking in injured and orphaned wildlife. My earliest memories are of a house full of animals, both wild and domestic–from the half-grown leopard who shared our home when I was two to the crows, ravens, foxes, and woodchucks–as well as cats, dogs, sheep, chickens, goats, and pigs–who came through our door over the years. We once even gave a Nile crocodile a short respite in our basement shower.

I often felt those animals were as much my teachers as the ones I had in school. Our own emotional communication is very similar to how most animals communicate. Yet some animals are more expressive than others. With both bears and humans, frowns are frowns and smiles are smiles. Bears, it turns out, are easy to read, once you begin spending large amounts of time with them. While my dyslexia prevented me from getting higher degrees, my sensitivity to emotion, extreme attention to detail, and capacity (which I long considered a drawback) to think in pictures rather than words have proven immensely useful in studying–and understanding–the behavior of animals.

Once I grew comfortable in my own study methods, I was able to understand how bears see, how they use scent and body language to communicate, how they compete and cooperate with one another, and under what conditions. I've been able to show that bears are social animals, not solitary as previously thought. And I've discovered an organ, now named the Kilham organ, which allows bears to identify plants and communicate with airborne scent. Moreover, I have been able to witness behaviors that more sterile views of animal behavior rarely recognize: altruism, reconciliation, and empathy, to name a few.

All of these discoveries have been enormously aided by working with juvenile animals, because—whether human or wild—all young animals have behavior that is more exaggerated than when they are older and more cautious. The cues provided by my string of cubs have been a tremendous help in decoding adult behavior. Yet beginning that work with juveniles was not an easy endeavor. Growing up in a house filled with wildlife and under the auspices of a naturalist father, both Phoebe (who had gone on to earn a doctorate in tropical soils) and I became licensed rehabilitators. Phoebe's interests lay in caring for animals—a task my mother, Jane Kilham, had excelled at—and I had hopes of eventually working with bears so that I could learn more about their ways. But it took two years before fish-and-game officials brought me my first cubs and extended me the permits to care for them.

Little did I know then that we would raise and return to the wild more than one hundred black bear cubs, or that I would establish a sixty-square-mile study area where I would follow and record the habits of wild bears. For many years, I was criticized by other scientists for not following traditional experimentation models. In fact, most of my observations that are new to science have gone unrecorded in official scientific literature. I enjoy the freedom that my independence offers, but without a doctorate or university ties, I also lack access to public funding and the ability to publish in scientific journals—both highly guarded by professional scientists.

I have, though, found comfort in the work of animal behaviorists like Konrad Lorenz, who wisely wrote in *On Aggression*, "Unless one understands the elements of a complete system as a whole, one cannot understand them at all." Even scientists like Niko Tinbergen, who along with Lorenz won a Nobel Prize, writes in *The Animal in Its World*, "It cannot be stressed too much in this age of respect for—one might almost say adoration of—the experiment, that critical precise and systematic observation is a valuable and indispensable scientific procedure which we cannot afford to neglect."

None of this means, of course, that I don't follow the scientific literature. In fact I scour it for anything related to my work. But I rely on the evidence I find in my own work, and use scientific literature only as reference. This has allowed me to forge not only new paths for rehabilitating bears, but also new methods of studying their lives.

Conventional science often gets hung up on debates over whether or not animals have emotions, are willing to make sacrifices to help others, or communicate with intent rather than out of instinct. This comes as a surprise, usually, to average people who have to look only as far as their dogs to answer such questions–at least as they relate to dogs. This is what leads me to say, often, that much of science seeks to prove and record what ordinary people already understand. That doesn't mean that it's not worth proving these things to be true, or that there aren't vast mysteries to be unlocked in the animal kingdom, but it does mean that new discoveries often feel like old news to the general public. In fact, I have been fortunate that my work has been received so well by the public at large–which has encountered it through articles, news reports, National Geographic documentaries, and at my own lectures.

And slowly but surely, scientists around the world are reaching out to learn what I've uncovered about bear behavior, and even seeking my advice on how to reintroduce bears whose populations are threatened. Ironically, too, while the typical science pathway was closed to me, my first book, *Among the Bears*, has been used by high school and college teachers to get students interested in modern science.

In reality, my methods are not much different from what was used by seventeenth- and eighteenth-century naturalists. After all, the work of somebody like Charles Darwin has stood the test of time, and his methods were much more similar to mine than to the scientific method used widely today.

But as much as I owe to thinkers like Darwin, Lorenz, and others who studied animals from the field, I owe even more to the bear who made many of my findings possible: Squirty.

CHAPTER TWO

Squirty's World

Lambert clearing is just a small corner of my 130-acre forested lot, but it is filled with clover patches, apple trees, berry bushes, and several small ponds and wallows–a perfect place for a bear to forage and relax. This is Squirty's inner sanctum, a place she often defends from other bears, and where she reigns supreme. I meet up with her regularly in the clearing, and it was here I discovered that she is part of a community–not only of mates and blood relations, but also of friends and rivals. But understanding her world took time, and her cooperation.

By the time Squirty arrived in my life, I had already raised Little Boy and Little Girl. It was March 1996, and a string of events that, I would learn, separated many young bears from their mothers brought her to my door.

Those events began the fall before, when seasonal rains forced loggers to stop their operations, leaving cut and bunched trees on the ground. To the loggers, this late-fall heap of unfinished business is just that–something that they must return to when the weather allows so that they can continue hauling away the timber. But to a bear those piles of wood and brush are something altogether different: They are near-perfect denning sites, with the settled brush providing shelter and the cavities between the logs creating chambers. So by the time loggers in black bear habitat return to their work sites in late winter or early spring, it's not unusual for bears to be holed up among the felled wood, often with cubs. Dogs could be trained to sniff out dens and avoid disturbing them, but the best solution is for the loggers to wait until summer–when bears have moved out of their dens–to go back on site. When loggers find dens accidentally, just giving some space will avoid problems. But the loggers who stumbled across Squirty's mother's den had taken no such precautions and were unwilling to leave it alone.

Her den was disturbed, and she fled. Attempts to catch her and reunite her with her three cubs failed. Eventually, the cubs, two females and one male, fell into the hands of wildlife officials, who brought them to my house in Lyme. Squirty was the smallest, weighing in at three pounds. The cubs, seven weeks old, quickly adjusted to life in a new den: a large basket in my guest bedroom. The cubs were being bottle-fed three to four times a day, and what went in one end came out the other, resulting in lots of laundry. They were too small and fragile to be outdoors. From the basket they were moved to a three-by-eight-foot enclosure with two-foot-high sides. It wasn't long before the male cub found his way out over the side and created quite a commotion after discovering that he was separated from his sisters. Soon we moved them to the basement, where I had built them a climbing structure out of ash logs to keep them from the structural beams–which had already received a workout from Little Boy and Little Girl, who'd also had a bit too much fun tearing out the room's insulation. By April, it was time to return to the forest.

As with the first cubs, I walked through the woods with Squirty and her siblings almost every day that spring and summer, helping them explore their natural environment and observing them as they learned how to be wild bears. At night, they returned to their protected enclosure, and as winter approached the cubs fattened up for their long nap and hibernated in a den I built for them. By the following June, they were finally ready to go off on their own. The male cub, whom I'd named The Boy, left the area, as did Squirty's sister, Curls. But Squirty stayed close, establishing her territory in the forest where she'd grown up.

By fitting her with a radio collar, I was able to keep track of where she denned that first year—and every year since—and reunite with her in the spring. Thus began our incredible relationship, which has allowed me to watch Squirty raise more than a dozen cubs, establish her home range, become the dominant bear in it, and form cooperative relationships with unrelated wild bears. She has been my greatest teacher.

Squirty's first set of cubs, Snowy and Bert, were born in January 1999, two years after she returned fully to the wild. During her first year with them, she pushed the boundaries of her home range. And as she did, I followed her, documenting where she traveled and the battle scars she developed along the way. Her contests with neighboring females were frequent, evidenced by the fighting wounds I found every couple of weeks when I measured her. Squirty looked upon me as her mother, so even though she was living as a wild bear, I could approach her easily, touch her, and measure her girth to get a sense of whether her natural food supply was abundant, or how much energy she was expending.

On a typical day, I would drive up the one town road that bordered Squirty's home range, trying to locate a strong telemetry signal. Once I knew about where she was, I would park off the road and follow her signal into the forest. By lowering the gain on my receiver, I could tell when I was close enough for Squirty to hear me and call out to her with my usual "Hi guys." Often Squirty would let me know she was near by breaking a stick, and sometimes she would circle out of sight behind me to pick up my back trail to be sure it was me. It wasn't unusual for her to make a soft repetitive "mm, mm, mm" on her approach to let me know she wasn't being aggressive. I always brought a small bag of kibble as a food reward to pay for the time I was taking from her. It was the amount of food my Labrador, Buddy, could inhale in less than a minute, but it took Squirty thirty-five minutes to eat it, as bears eat their food one piece at a time. This gave me time to measure her and check her up close for injuries.

Those injuries all turned out to be only minor scratches, but they were indicative of face-to-face encounters with other bears. I assumed that all these contests were with unrelated resident females, as she had no adult offspring. It was too early in my research for me to know who these females were, or even if there was more than one. But it was clear what they were fighting over–a summer food supply in a recently logged, hundred-acre lot that was now in the berry stage.

I observed and filmed the end stage of one of these meetings, with a sow and two cubs. I had been tracking Squirty by her collar signal, and when I came upon her she was treed with one of her cubs above her and one up a separate tree. The area's resident female, clearly the aggressor, was leaving with her own two cubs in tow as I arrived. Squirty was highly disturbed, chomping and huffing, and she remained in the tree for several hours after the resident sow left.

It wasn't until years later that I would learn that territorial scuffs between sows were not by chance. Yoda was another cub I had raised and released. One day, after she had made a life in the wild, I was walking with Yoda and her cubs and twice watched her pick up the scent of another bear who had been feeding in what she considered her home range. First, she back-rubbed over their marks; then she tracked them step-by-step through the forest, leaving minute drops of urine that dripped from her marking hair–a specialized tuft, about four inches long, that surrounds a female bear's vulva and allows urine drops to flow down it, past her fur, and reach the marking target. (This tuft of hair can be useful in identifying a bear as a female.) Urine by itself gives no scent signature, but each hair has both sweat and sebaceous glands that infuse the urine with scent. I never was able to stay with her long enough to see her catch up with her victim. Either night would begin to fall and I'd have to head for home, or she would turn and false-charge—essentially asking me to leave to prevent my interference.

It was clear through all of these observations that the bears had home ranges that overlapped at major food sources. It was also clear that they were possessive of the resources they needed to survive. They stood their ground to defend them, but over time and repeated encounters, their aggression would lessen and they would slowly build alliances with other bears. There is no way of knowing how often these alliances form, but I suspect it is common due to bears' ability to travel outside their home ranges to access food in the fall shuffle. Documented by many researchers, this shuffle is a time when bears travel short and long distances to feed on abundant surplus foods when the food supplies in their home ranges have waned.

And so I would watch, over the years, complex alliances emerge among Squirty and other resident females in Squirty's home turf, and my study area. Bears, as I've learned from the intimate experience of raising four sets of them as a surrogate mother, are assertive and even aggressive in getting their own way, but at the end of every conflict I have had over the years, there has been forgiveness. When it was over, it was over. I learned this early on, with my first set of cubs. When I started walking with LB and LG, I would coax them into dog carriers and drive to new places for interesting excursions. After one of the walks, LB refused to get back into the carrier, so in frustration I tried to stuff him in. That didn't work very well: He exploded, roaring and lunging at me from a bipedal stance. I finally calmed him down by pushing him over with my foot, but we were still two miles from home and that meant a long walk back. I didn't calm down as quickly as he did, and he spent the whole way back at my heels making the soft repetitive "mmm, mmm, mmm" moan of reconciliation. He was upset because I hadn't forgiven him. Typically cubs have ferocious fights over food, with roaring, biting, and fur flying, but seconds later they are fine with each other.

I now realize—after witnessing over a thousand social interactions and being involved in many myself—that aggression in bears can be and often is a stepping-stone to friendship. Friendships and alliances frequently develop by repeated interactions, with initial aggression that lessens over time. There is also aggression between friends as the limits of interactions are established. The bears who allow me to enter their world and observe them signal their limits to me in the same way, and understanding and accepting those signals has been crucial to my work.

Some bears in my study area, like Squirty, have accepted me as part of their family. I have raised them or their mothers, and so I have been granted special permission to enter their world–as long as I follow their rules. Other bears I track are wild, unrelated animals I have trapped and fitted with telemetry and GPS collars–something I've been doing since 2003 when I began a cooperative study with the New Hampshire Fish and Game Department. I was interested in expanding my bear behavior studies, and the NHFG was interested in finding out more about cub survival so they could better project bear populations in the state. The plan was to collar ten female bears in my study area, which is sixty square miles in the eastern half of Lyme and the western half of Dorchester, New Hampshire. The bears would be a mix of Squirty's relatives and females unrelated to Squirty. Being able to track these bears' travels, as well as observe them routinely in the wild, has revealed much about the way bears define and defend their turf, interact with family members, competitors, and mates, and–most surprising–negotiate the social contracts that define their lives. We currently have working collars on eight bears.

The adult female bear's home range is as close to a territory as exists in the bear world. Female bears must carve out a home range of high enough quality to support the mother and her cubs while the cubs have a limited ability to travel. Squirty and other females will push to expand their boundaries. Not all cubs will be able to stay in their mother's home range as they mature, but it is an advantage to female bears to have daughters to fill and expand existing home ranges. So if conditions are right (bears don't share the resources they need to survive, only the surplus), mothers will make room for female offspring–and in the process build what I call a greater home range.

Squirty has continually expanded her home range through contests with other bears. Contests between females usually involve the dominant female chasing or treeing the subordinate. Occasionally two females will engage in face-to-face spats, striking out at each other with their claws and vocalizing with the negative "huh, huh, huh." But Squirty has also expanded her range by taking advantage of opportunities when neighboring females have been killed in the hunting season. In years when she has surplus home range for a daughter to fill, she will nurse her cubs a full eighteen months. Only one female will get to stay, though, and if there is more than one daughter it's not Squirty who picks which one that will be. The dominant daughter will chase off the others. That was the case when Squirty's daughter SQ2, born in 2003, drove away her two sisters. When Squirty's first daughter, Snowy, had a trio of trio of female cubs in 2004, her daughter SNLO (short for "Snowy's Little One") drove off her two sisters. But when Snowy was killed in 2007, SNLO's second sister, Two, was able to take over her mother's home range. So it is in this way that Squirty's daughters and granddaughters who live in the greater home range continue to expand their portions of it, and evolve their family group.

In years when there is no new real estate, Squirty will stop nursing her cubs in the winter den. To understand why, you need to understand a bit about the bear breeding cycle and the factors that drive it. Bears pick the spring of the year for their breeding season because that's when food is most evenly dispersed across the landscape, and females need six months without nursing cubs to get fat enough to reproduce. Females are in estrus from roughly the last week of May to the first week of July. During that time males will advertise their availability to mate, and a female will seek out the male who sparks her interest. Eventually she'll choose a mate, and they'll spend three to seven days together.

The result of that union will be a two-celled organism called a blastocyst. Since bears are delayed implanters, the blastocyst won't embed itself in the uterus until late November or early December, after the bears enter their dens. There's a short gestation period of fifty to fifty-five days, and cubs are born in the dead of winter, in January, weighing less than a pound. Their eyes are shut, their little earflaps are down, and they continue to develop outside the womb while nursing in the warmth of their mother's fur, in the dry cavern that will shelter them until spring. They are not even able to follow their mother until she emerges from the den for good in April.

When she does, the first thing she will do is build a nest at the bottom of a good climbing tree and coach her cubs as they climb up it for the very first time. The cubs have to learn how to safely traverse rough bark, smooth bark, dead bark, and the skinny end of a limb. Not until the cubs are adept climbers will their mother be able to travel with them, moving them from babysitting tree to babysitting tree–where they will sleep high up in branches as she forages for food. Like a human mother listening for a cry from a napping baby, she'll be alert to her own particular clue for danger: the sound of a cub's toenails on bark. When a mother bear hears that from her babysitting tree, she responds immediately. Around this time, the mother bear will begin to train her cubs when to follow and when to stay. I have watched Squirty turn and false-charge her cubs, sending them back up the babysitting tree when she didn't want them to follow her. In time she developed well-honed control over each of them, something I was never able to accomplish while I raised and walked cubs.

Throughout the summer, the mother helps her young explore their world, and they follow her back into the den in the fall. They emerge the next spring as yearlings, and soon the mother will again go into estrus. When the chosen male bear shows up, the family unit will break up for the three to seven days that the mother and father travel together and mate. When she returns, the yearlings react to her as if everything is normal–greeting her in an "Oh, great, Mom's back" kind of way. But when they get within about twenty or thirty yards of her, she'll chase them and run them up a tree. The message: Things aren't normal anymore. Once she has bred, she can no longer nurse, as it could interrupt her pregnancy. And so the separation begins. In a year when there is space to be had in her range, and she has at least one daughter, she will nurse as long as possible before mating to keep her cubs close and let a daughter secure a spot in her domain.

Watching Squirty practice her motherly tough love, I've come to understand that when she stops nursing, she begins to weaken the bond between herself and the cubs, making it easier to keep them from staying around. I've seen her start to chase her cubs at just fourteen months to ensure they weaned early and vacated her turf.

The core area of the typical female home range is three to five square miles, with some overlap into neighboring females' home ranges. But home range sizes vary from place to place across the country; they depend entirely on the quality of the habitat. More food in a smaller space means a bear doesn't need to carve out as large a territory. A sparser food supply, on the other hand, means a bear needs more territory to survive. Single subadult males, generally under eight years old, are chased out of female territories, but appear to access surplus foods by forming groups of friends and overpowering resident females. Adult males are asked by females to leave and not compete with them for food and other resources–such as space, cover, and water. The fact that they comply signifies a degree of female choice in mating. Breeding males are much larger than females and could easily stay in a female's turf and compete with her for food, but they choose not to because of the repercussions on the mating front.

Overlapping these female home ranges will be the home ranges of several breeding males. One dominant male may mate with multiple females in his breeding area–introducing another level of relatedness among bears in any given territory. In 2011 a single male, who arrived in my study area on June 16, halfway through the mating season, mated with Squirty, SQ2, and SNLO (all part of Squirty's clan, with whom she shares her greater home range), as well as Moose and SN2 from the neighboring clan, in which Moose is the dominant female. Due to his input, the offspring of both Squirty's clan and Moose's clan will be half brothers and sisters, and over time the resident population of females will become more and more related. There's no risk to genetic health: Young males move away, and Squirty has mated with nine different males so far.

The lives of male bears are decidedly different. They roam widely and feed on abundant surplus foods where available. It is not unusual for a male bear to travel over a two-hundred-square-mile area in a single year. They live transient lives, and the population of males in any given area changes over time.

Males begin their itinerant ways as yearlings, after their mothers–and aunts and grandmothers sharing their ranges–begin to relentlessly push them out over the spring and summer. By September, they are on their own, often befriending other young males for better access to food and other resources. It will be some time before they mate; any thoughts along those lines will be quickly subdued and suppressed by the individuals (10 to 15 percent of the male population) who are big enough to compete for females. Subadult males do, however, sometimes mate with subadult females who are of no interest to the adult breeding males, as they rarely are able to raise their young.

Over the years, the cast of characters coming in and out of Squirty's range has grown large, as has her family tree. She now shares her inner sanctum at the Lambert clearing with members of her own clan, with Moose (who is at least ten years older than Squirty), and with all of Moose's clan. There are conditions for sharing–as well as limits and reasons for it. Interestingly, it appears that many of the females now friendly with Squirty and her clan are the same ones Squirty once competed with for resources as she initially fought for her own space to raise cubs and prosper.

And so it is that by letting me watch her at and around Lambert clearing, Squirty has revealed to me the family lives of bears and introduced me to bear behavior that would typically take place far from human view.

Every set of cubs that Squirty had brought surprises. In 2005, for instance, she gave birth to two female cubs, but had no new territory to place them in. She not only had broken off nursing in the winter den the following year, but when she returned to the clearing in April 2006, one was already gone and she began chasing the other to encourage the family to break up early. She hooked up with Burt, her longtime male friend, and spent two months with him before mating when she came into estrus in June.

In January 2007 Brooke, a female, and Grant, a male, were born. Grant left the area in September of his second year, and Brooke stayed in the area, showing up at the clearing after Squirty went into her den to hibernate. She gave birth at three years old to a single cub. It's not unusual for first-time mothers to fail at raising their first cub, and that's what happened to Brooke. Sometimes new mothers can't put on adequate weight in the fall and simply run out of milk for the cub in the spring. Once they stop lactating, they automatically come into estrus. If the cub is still alive when the breeding season starts in late May, it may be abandoned when the male shows up to mate. However, the mother's lack of experience may also have something to do with a cub's survival. Brooke was unable to secure a home range or give birth to cubs in the two subsequent years, but she was able to stay in the area because she formed a friendship with a subadult male. A coalition of two is more powerful than one, so by traveling and feeding together, they could avoid being chased out by resident females. I was able to document their relationship, catching them feeding side by side in front of monitoring cameras I had set up throughout my study area.

Eventually, though, my tracking of Brooke taught me as much about people as it did about bears. She was killed at a bait site—a spot where a hunter places a pile of food, usually stale doughnuts, to attract bears. The hunter sits and waits until the bear arrives, and he or she shoots. It's a practice that's illegal in many states, but some hunters still defend it, claiming that observing the bear at the bait trap helps them avoid shooting sows with cubs, or tagged or collared animals. Brooke had dropped her radio collar, but she still had ear tags; unfortunately they did not help her in this situation. I spoke with the hunter, who was very remorseful about shooting one of my research animals, and he told me she was with another bear, presumably her male friend, at the time of her death.

When the spring of 2009 rolled around, Squirty had just given birth to a single male, Cubby. I was planning on a summer of watching the interaction between Cubby and his mother. When Squirty has two or more cubs, the cubs play with each other and only occasionally with her. Now there was just one cub, and I wanted to see how the dynamic changed. But things don't always go as expected, and in the beginning of June I saw Squirty initiate nursing, then almost immediately break it off. The next day I witnessed an equally unusual event, as Squirty took off after SQ2–her daughter and the number two female in her range–in a hierarchical chase with Cubby right behind her. That struck me as odd: In all my years of watching her, she usually safely put her cubs up a tree before taking such aggressive action. By the middle of June, Squirty and Cubby left the clearing, and I did not see them again for a week. But all during that week I did see a number of large males on my monitoring cameras at the clearing.

By the time Squirty returned, Cubby was gone. It was clear that Squirty had mated, and the series of odd events began to make sense. She had intentionally stopped nursing Cubby, triggering estrus and ovulation. This being a novel act, it is likely that Squirty recognized a pattern in her own reproductive behavior and knew that she needed to stop nursing to induce estrus. In other words, she planned the event, projecting she would have a better result in the future. In the world of bears, Squirty was being a survivor–essentially employing the same kind of techniques used in captive-breeding programs to maximize the production of offspring. It seems that Squirty is well aware of how to maximize her lifetime productivity, and I suspect that she values female offspring more than males: They are instrumental in building greater home ranges, while the male offspring leave and are rarely seen again.

If there's one thing I can't help being, though, it is human, and I thought the worst for Cubby. When female bears come into estrus, the condition goes out on the airwaves, automatically attracting males and exposing any living cubs to males who may not have their best interests at heart. Over the years, I haven't seen any cubs killed by males attempting to mate with their mothers, but we have had one rehab cub come in with a bite from a male. Three weeks passed and I thought Cubby was a goner. Then one day there he was, following Squirty into the clearing. He was in good order, hungry, and had a short scar between his eyes. Things were different now, though: He could follow Squirty, but if he approached within six feet of her, she would turn and false-charge and roar at him.

Over the summer, Cubby stayed in close pursuit. It was a win–win for Cubby: He wanted to be close to Squirty, but if that didn't work she would leave to avoid him, and Cubby would get her food. An interesting twist to all of this was that Cubby maintained his social status as Squirty's cub and could take food from SQ2 and his adult female cousins. Did this mean Squirty was still defending him when he was challenged by other bears? And if so, why was she? Or was it simply fear of reprisal based on her rank? Normally if a cub's mother is killed and the cub is orphaned, the cub loses all social status–there is no living bear with the power to back it up. But Cubby was tolerated, even though a standard reading of the bear world would lead one to believe that wasn't possible. Clearly, there is far more to learn about the emotional lives of bears, and their penchant for empathy.

The last I saw of Cubby in 2009 was when he followed Squirty from the clearing as she left for the year to den. I didn't see him at the clearing again until the fall of 2010, and he was smaller than Squirty's new cubs but acted like a two-year-old bear.

The biggest surprise, however, began to unfold in 2010, and it, too, showed me that bears can act out of empathy when doing so doesn't put their own survival at risk.

It was April, and I was hiking two miles toward the base of Smarts Mountain to visit Squirty's den. I had hoped to retrieve her collar, as she has a habit of removing them and leaving them in the den. Instead, I found her up a tree with a single small cub. It would be another two weeks before Squirty came down off the mountain. When I caught up with her she was on a neighbor's land, at the site of her first maternal den. It was an area that had been lightly logged more than ten years ago, letting sunshine in on the forest floor and allowing a rich growth of sedge and broad-leaved herbs, or forbs. There were some wetlands, and an ample scattering of large white pines that served as babysitting trees–where she could tree her young while foraging or when an intruder was in the area. When I left, Squirty apparently followed, as she was at the clearing the next evening.

Meanwhile, Squirty's daughter SQ2 had given birth to three cubs (two males and a female) in a den near Pout Pond in Lyme, near a source of acorns left over from the fall crop. Lyme is a small town with neighbors who know one another well, and many take an interest in the bears. So I was getting regular updates. It appeared that SQ2 was spending a lot of her time in the oak stand to the east of Put and Marion Blodgett's house on Flint Hill. She and the three cubs fed on green vegetation in Put's field and were seen passing through Lee and Kathy Larson's field on their way to a clover patch at Ben and Mary Sue Henzey's. Early in May, Put called and was concerned that two of the cubs were very active and always played together while the third cub was smaller, often apart and sleeping in a tree. I explained to him that it was not unusual for a sow with three or even two cubs to have one quite smaller–or even lose one–as a result of cub competition for milk.

Life at Lambert clearing was bustling. Squirty was there with her male cub MC4. Unless MC4 was high in a large tree sleeping in a cradle of limbs, he was in Squirty's shadow. While Squirty ate, he would take refuge between her legs, sometimes lying on his back and sneaking some milk and at other times poking his little head out from between her front legs.

Several of Squirty's clan had returned to the clearing, too. First was her granddaughter Two with her yearling daughter. This was a new year and a time for a new understanding of Squirty's rules. Squirty stashed MC4 up a large pine and proceeded to charge Two, treeing both Two and her daughter up another large pine. Then with a series of stiff-legged walks and full back-rubs against trees, as well as by walking over saplings to spread her scent, Squirty made it clear that she was still the boss.

The next to arrive was SQ2LO, SQ2's daughter from a prior set of cubs. Two had treed, showing respect and submission, but this wasn't the case with SQ2LO. She took off with Squirty in close pursuit. It was close to ten minutes before Squirty returned to the clearing. Granddaughters, it turns out, understand their mother's rules, but sometimes have a much harder time understanding that they have to learn and obey their grandmother's rules as well. Spring was in full swing. I was visiting the clearing daily. I had visited each of my study animals over the winter, changed their GPS collars, and was tracking their activity.

Then one Wednesday in May, as I was leaving the clearing, I encountered SQ2 waiting for me by the trail. She was timid and engaging; it was as if she had something to say, but I wasn't getting it. I gave her a treat (something she'd grown accustomed to, as Squirty's daughter) and left for home. It was the first time I had seen her at the clearing that year. The next day I returned to the clearing early but did not stay as I had to lecture that evening. When I got home, I had a message from Beverly Balch that there was a cub bawling relentlessly in a large pine at the bottom of their field and that SQ2 had been there with her cubs that morning. They were going out for the evening. They called back when they returned several hours later to say the cub was gone. We both assumed that SQ2 had come back for the lost cub. I would later figure out that the cub wasn't lost at all, just confused by the events that were transpiring. I would also figure out that SQ2 was trying to convey to me that she had a problem and was concerned about her daughter's survival. I wished I could have captured her expression and demeanor on camera–she was clearly trying to communicate with me.

But I didn't piece together the puzzle until Saturday evening. Squirty was waiting for me at the clearing, which by itself was not unusual, and I noticed MC4 sleeping up in one of the large pines. I didn't think much about it. While I was observing the bears who had shown up, a large breeding male followed Two's trail into the clearing, sending her up the big pine with her yearling cub. The male, who happened to be MC4's father, checked out scent that Two had left, then rubbed his back against a maple tree, acted out a stiff-legged walk, and walked over saplings to leave his mark and signal his intent to mate with her. The commotion sent Squirty gulping and running to her babysitting tree to prompt her cub to climb higher.

I looked up with great surprise as I heard two sets of claws scratching their way up the pine. There were two cubs! Squirty had rescued and adopted Josie, SQ2's weakened cub.

What I saw that day raised huge questions about cognition, empathy, planning, and communication in the bear world. It was clear that the adoption was deliberate, and the GPS data tracking their whereabouts indicated that the terms between Squirty and SQ2 were negotiated in one evening. SQ2's expression and demeanor on what turned out to be her first trip to the clearing, along with testimony from the neighbors about her behavior with her cubs, supported the notion that SQ2 recognized she had a problem. Her decision to go to the clearing and her knowledge that Squirty would be there suggests a planned event. The fact that the adoption took place in a single evening and was enforced over the ensuing year suggests that SQ2 and Squirty could negotiate the terms of a complex agreement without language. SQ2 recognized Josie's plight, knew from past experience that Squirty could be the solution, and was able to both relate to the past and project into the future, a recursive process thought to exist only in humans.

Gifts like this observation have come from following so many bears from cubs to adulthood. I've been able to see behaviors that I had witnessed in my orphaned cubs replayed in the wild, in "real-life" situations. And of course I've been able to watch Squirty, now seventeen years old, live out her life. Events like her adoption of a cub in need show there's much I still have to learn from her, and I can only hope she'll be around long enough to teach me.

Though I have learned from them all, I haven't been as lucky with every cub I've raised. Yoda let me enter her world so fully that I could literally hold my video camera between her front paws and film her tongue licking up ants and ant larvae as she bit into a log or a stump. I often think of what I might have witnessed in her world, and her offsprings' world, had she survived beyond five and a half years old. But like many bears, Yoda was drawn to remote uninhabited cabins that contained food—a trait that often leads bears to untimely deaths.

I kept track of her activities and consulted with the cabin owners, all of whom were extremely cooperative. But there were some hunters who had access to the twenty-two-thousand-acre property on which Yoda lived. The owner allowed me to conduct my research there, and also allowed these hunters to hunt there, on the condition that Yoda was not to be shot. To be sure that there would be no mistakes, he asked the hunters not to take any bears in the portion of the property that was the center of her home range.

The hunters agreed, but apparently had other plans. One early-November day, they shot her within ten feet of her winter den—the property caretaker and I were able to reconstruct the events that led up to her death by following the hunter's tracks in the fresh inch of snow that had fallen the previous night. When she was shot, Yoda was wearing a telemetry collar with four orange three-by-five-inch cattle tags, and the hunters had walked by signs stating COLLARED BEARS IN THIS AREA—PLEASE DON'T SHOOT.

I'm sure those hunters will never understand the gifts of knowledge that Yoda could have lent us as she revealed the inner workings of her world. And I know that most hunters act with more respect. But the loss of Yoda makes me ever aware that my time with Squirty, the longest-living bear I have raised from a small cub, is both precious and rare.

Chapter Three

Bear Society:

Beginning to See the Forest through the Trees

When Squirty was ten, I began to bring her food with nightly visits to the clearing—about a gallon or so of corn, something that had about the nutritive value and digestibility of acorns. The mast crop had been poor for two years in a row, bears in the area were turning to backyards for food, and I needed to come up with a way to keep her from frequenting the house of a rankled neighbor. He had previously, and unfortunately, been putting food out to attract Squirty to his yard. And while he still had food outside, apparently for other reasons, he was threatening to shoot her if she came back to feed. When food is left outside homes bordering their habitat, bears don't know the difference between a WELCOME mat and a NO TRESPASSING sign. So I decided that giving her food at her home site would keep her from wandering into dangerous territory. It worked, and I quickly learned that offering a small food reward was a good way to see bear behavior and social interactions. Sometimes those interactions were quite unexpected.

Bears live in a world of feast and famine. One summer, they might feast on a huge raspberry crop in a clear-cut; the next summer, there may be none. The trees they live among might produce an enormous nut crop one year, but only a few nuts the next. The amount and reliability of their food supply determines not only their survival, but also the survival of the next generation. In good food years, both dominant and subdominant females can raise cubs, but in poor food years, or places with a great deal of competition, a female might reach the age of five or six before raising her first set of cubs.

The vast majority, about 85 percent, of a bear's diet is vegetative—plants, berries, roots, and nuts. In the early spring, when the bears' food is most evenly dispersed in the landscape, they feast on the buds of beech, oak, and ash. They munch the flowers of white ash and red maple, and the leaf starts of beech, poplar, and oak. Along woodland trails they find nodding sedge, jewelweed, and wild lettuce. In years when nut crops fail, they rely heavily on jack-in-the-pulpit, their major root crop. By July, food is primarily available in patches: That's when bears rely on spots where raspberries, blackberries, and blueberries grow, and travel to wetland swamps to find blueberries, huckleberries, mountain holly, nannyberries, dogwood, and cranberries. By fall, they must look for food in stands of red oak and beech. Ants, bees, and grubs provide most of their small amount of protein, but if they have an opportunity they'll also take newborn deer fawns and moose calves, young birds from nests, and will sometimes grab a fledgling bird off a branch—or, on rare occasions, take domestic animals like sheep, goats, chickens, or rabbits, all slow versions of natural prey.

Because their food supplies vary so much in space and time, no one bear, or even group of bears, could defend a territory large enough to provide all the food required for survival and reproduction. To secure the food they need, Squirty and other female bears maintain overlapping home ranges, developing long-term cooperative relationships with other individual bears.

Moose, the oldest bear in my study, was captured, ear-tagged, and collared in 2004 as part of research I undertook with the New Hampshire Fish and Game Department. She was twenty-three at the time. Her home range is dominated by beech trees, which theoretically produce nuts every other year but are often unreliable. Squirty's home range, in contrast, is composed largely of red oak stands, which have abundant nut crops only once every four to five years, with intermittent crops—and sometimes no crop at all—in between.

One October day in 2005—a year when Squirty's oak stands produced a large nut crop while Moose's beech produced very few beechnuts—I visited Squirty's clearing and put two small piles of corn on the ground, fifty feet apart. Squirty, who was there when I arrived, was soon feeding from one pile. Moose came down the hill, passed about ten feet behind Squirty, and began feeding at the second pile of corn. To my great surprise, Squirty ignored the unrelated bear and calmly continued feeding. Why didn't Squirty just chase this invader out of her treasured clearing?

It was then that I realized that Moose has a cooperative relationship with Squirty, and with Squirty's clan of relatives. Since that first sighting of Moose in the clearing, I have documented her many visits to Squirty's territory. Initially Moose appeared at the clearing without cubs, most often visiting when she needed to fatten up to give birth. Now, apparently more comfortable with her relationship with Squirty, she routinely brings her cubs to the clearing in years when there are plenty of acorns, but rarely in beechnut years. With rare exceptions, Squirty willingly shares her extra food. And Moose is far from the only bear to share Squirty's bounty. One year during a widespread natural food shortage, Squirty shared her resources at the clearing with at least twenty other bears, many of them strangers. Remote cameras captured all the action. Squirty isn't the only bear I've seen share resources. One fall we set up a meeting place for Yoda. We left food there and monitored it with a remote camera. On numerous occasions, we saw Yoda share her food with one bear, and her twenty-acre clear-cut with another.

Likewise, data from Squirty's radio collar has shown her feeding in other bears' territories, and ongoing research using GPS data is revealing new relationships and resource-sharing arrangements among the bears in my study area. These relationships are mutually beneficial. Since the bears' food sources are all unevenly distributed across the landscape for most of the year, one bear can enjoy a surplus in its home range while another suffers a famine. The following year, their situations may reverse. Sharing, therefore, is a matter of survival. During my years of observing Squirty and other bears in the wild, I've seen them develop and enforce a complex array of these social contracts. The more alliances a bear forms with its neighbors, it appears, the more likely it is to secure food in times of shortage. These social contracts minimize the amount of energy expended in aggressive interactions while gaining access to surplus foods.

When we think of social animals–that is, animals who live together in well-defined groups, and form enduring relationships–we usually think of the great apes, of wolves and other members of the dog family, and, of course, of humans. Science considers bears to be solitary animals. But while bears don't live in established groups or obey rigid hierarchies as chimps or wolves do, they have amazingly complex social relationships. Each individual's behavior is controlled by all of the individuals with whom it shares resources, and shifting hierarchies form within families, among males, and among unrelated bears at any place that bears congregate to feed. They communicate through facial expressions, ear movements, body posture, and scent cues. They also maintain their relationships through judgment and punishment–behaviors thought to exist only in humans. Bears, in short, are part of sophisticated societies that we are only beginning to understand.

As I witnessed the bears in my study area over time, I began to see the forest through the trees, to understand the alliances that were formed, and why. I also learned that, for a bear, successfully navigating social waters requires an early start. Each cub needs to develop a personal coalition of other bears willing to share the resources required to survive and reproduce. With persistence and fortitude, a newly independent young bear carves out its own place in society–even, at times, standing up to its own parents or grandparents, as one of Squirty's granddaughters did to her one spring.

At the time, Squirty and a bear named Burt were inseparable. They came to the clearing together in the evening, and spent the rest of the time on the hill above, eating acorns and sleeping. Beginning when he was three years old, Burt would come to the clearing once a year, just after breeding season. Each time he arrived, Squirty would pay special attention to him. Once, I stood over a small amount of food that I'd put down for her, and she came over and shoved me out of the way so that Burt could eat. This time, Burt was eight and a half years old and ready to breed. Meanwhile, Squirty had driven off her cubs early–in April rather than June. She spent the rest of the spring frolicking with Burt.

Another regular visitor to the clearing was one of Squirty's granddaughters, Snowy's Little One. One of a set of identical female triplets, SNLO had just turned two. She was old enough to breed and, if the fall nut crop was good, would have cubs in the winter.

The previous year, when the triplets were small and food was scarce, Squirty had shown no aggression toward SNLO or her other granddaughters. But when SNLO, the dominant cub of the trio, chased her siblings away from the clearing, Squirty's tolerant attitude changed.

SNLO was, then, a lanky, thin eighty-pound bear. Over the next three months, Squirty continually stalked, chased, and treed her granddaughter. Immediate access to food did not seem to be the issue. There was plenty of it to go around, including an ample supply of leftover acorns from the previous fall. And it wasn't just that she didn't want to share. Squirty would eat side by side with Burt, sharing her food, and when Snowy or SQ2–both Squirty's daughters–visited with their respective cubs, Squirty would give them a short dominance chase, but let them stay. She was also perfectly friendly with me, even after aggressively chasing SNLO.

It was clear that SNLO was being singled out. But why? I suspect Squirty recognized that SNLO was laying claim to her territory, and knew that in order to protect her resources for the long term she would have to make sure SNLO understood her rules. Squirty was boss, after all, and everyone down the chain had to understand that. She had made a judgment, and was enforcing it.

One afternoon, I watched Squirty locate SNLO by following a downwind scent, and, with the stealth of a big cat, stalk the smaller bear. With her ears turned toward SNLO, tracking any movement, Squirty crept forward in absolute silence, carefully placing each step. Periodically, she stopped in mid-stride, a scene reminiscent of my Brittany spaniels on point. With the help of good cover, she came within about twenty feet of the unsuspecting SNLO. Squirty charged, and SNLO, younger and quicker on her feet, took off. After a short, spirited chase of about two hundred yards, SNLO sought refuge in a large white pine tree, and Squirty climbed up right behind her. I ran toward the sound of breaking limbs and claws tearing into bark, hoping to photograph the clash.

Both bears raced about ninety feet up into the living branches, and SNLO took refuge fifteen feet out on a limb. Neither of these bears could do equations any better than I could, but they both understood the physics of the situation: The limb would support one of them, but not both. Squirty voiced her displeasure with a series of chomps, huffs, and a scornful and reverberating "HUH, HUH, HUH." As an exclamation point, she pulled and broke all of the smaller limbs around her.

After about five minutes, Squirty began to descend the tree, but this, it appeared, was just a ploy to lure SNLO off the limb. As SNLO followed her, Squirty made a second assault, catching SNLO on a thin branch. When the young bear leapt to a larger branch, Squirty's teeth sank into her side. Things looked bad. A ninety-foot fall could kill a bear. But when the flurry of teeth, claws, and reverberating vocalizations ended, SNLO was literally out on another limb–and I had a more heartfelt appreciation of the expression.

It seemed that Squirty had done what she could to teach SNLO that in order to share and benefit from Squirty's surplus, she would have to honor and obey the boss. At the bottom of the tree, Squirty paced back and forth, using a stiff-legged walk, and marked the ground with drops of urine–another reminder of who was in charge. Burt came over to inspect the commotion, carefully checking all the signs.

This game was by no means over. SNLO was persistent, and even brazen enough to taunt Squirty into chasing her. I witnessed more than a dozen chases over a period of two months that summer, and I knew many more took place out of my sight. SNLO was naive: Nothing in her experience as Snowy's daughter had taught her that her grandmother had house rules, and that SNLO had to obey them. Just like human children, who face a separate set of rules when they visit their grandparents, SNLO needed to learn to live with a new authority.

Unfortunately, I had more of a hand than I would have liked in the final stage of the Squirty–SNLO saga. I wanted to put a radio collar on SNLO, but worried about what might happen if Squirty arrived and found her in a weakened condition. Still, I was running out of time. I needed to stop capturing and sedating bears by the first of August to allow time for the drugs to completely metabolize before the beginning of the hunting season. So one afternoon I sedated SNLO and fitted her with a collar and ear tag. I worked quickly, also taking some measurements and collecting hair and oil samples for analysis. I had given her a light dose of the tranquilizer in the hope that she would awaken quickly–which she did, stumbling off into the bushes. Just then I looked down over the hill, and there was Squirty.

Things did not go well. I tried to distract Squirty, but she was far too angry and blazed past me toward where SNLO was trying to hide. I raced after her, yelling–to no avail. Squirty was already on the attack when I caught up, standing bipedal with her head down, ripping off one horrifying reverberation of "HUH, HUH, HUH" after another. SNLO simply didn't have the energy to fend off Squirty, who charged forward repeatedly, attacking with a furious ensemble of teeth, claws, and deafening vocalizations. SNLO just flopped around helplessly. Finally, SNLO managed to get away and climb partway up a tree. To my relief, Squirty didn't follow, but turned back toward the clearing, allowing SNLO to slink away.

I didn't see SNLO for four weeks and was worried about her. I was able to monitor her movements via the collar; they seemed to be normal. But the fact that I had played a role in her inability to fend off Squirty's attack weighed on me. It is impossible to study the behavior of any animals without interfering on some level, and possibly injuring them when doing so. In all my years of working with bears, though, none has sustained a major injury because of my research, and I didn't want SNLO to be the first. I take some comfort in the fact that routine interactions between bears are more rigorous than anything I do to them–like changing a radio collar, or tranquilizing them in their dens in order to take their measurements, remove hair samples for DNA analysis, and, occasionally, remove a tiny premolar tooth to age the bear. I've also learned that, just as bears forgive one another, they forgive me.

When she finally returned, she seemed okay–and from then on, she followed Squirty's rules and was treated in return as a full member of the group. Today SNLO is raising her second set of cubs and has become a trusted partner in Squirty's hierarchy.

Because they travel widely, often over more than one hundred square miles, male bears are much harder to study than females. And since male cubs are usually pushed out of their mothers' territories, it's more difficult to maintain close relationships with them. But in March 2008, I was able to collar Burt, who was then ten years old, and observe his social relationships during the breeding season and beyond.

At the time, Burt was not fat, but fit, weighing 226 pounds and measuring seventy-two inches from tip of nose to tip of tail. He had the five-inch-wide front foot pad of a mature male. From the time he left his den in early April until the middle of May, the beginning of the mating season, he stayed within telemetry range, in the forest tracts and rural developments of southwestern Lyme and northwestern Hanover. His range had plenty of good food that year, with some leftovers from the fall nut crops and a spring crop of black-oil sunflower seeds, which have twice the calories per unit of acorns. Burt also took advantage of rural human residents. During the season when people are not supposed to feed birds, I received five different reports of him at feeders.

When the mating season started in mid-May, he began living like an itinerant country gentleman, traversing most of Lyme and parts of nearby Orford, even making excursions across the river into Vermont, to visit prospective mates. During this period, it was not unusual for him to travel ten to fifteen miles a day, and I got a lot of calls from people who had seen Burt crossing roads, passing through yards, and traveling with mates. He successfully mated with at least two females, one on Bear Hill in Lyme Center and another near Mud Turtle Pond in Orford, about six miles away. He also made himself available to at least three other females in Lyme and in East Thetford, Vermont.

Burt's mates, I learned, got more from the relationship than a litter of cubs. During the summer, I received three calls describing a "female bear with a very healthy cub" eating from bird feeders. When I saw the pictures, I realized the bears were not mother and cub, but Burt and a much smaller mate. The females, it appeared, were taking advantage of Burt's "protection" to forage in places too risky to explore on their own. One spring day, I was interviewing a prospective summer intern when Cole2, a female who had mated with Burt, climbed up on our porch in broad daylight, searching for a bird feeder. Burt, it turned out, was waiting for her a few yards away, and later accompanied her as she walked past the sugarhouse and barn.

I have seen this "bodyguard" behavior on other occasions. Squirty had made it clear on several occasions that Yoda was not welcome in the clearing. Yet while traveling with a very large mate, Yoda returned there, parading past Squirty and her cubs while Squirty silently observed them. Squirty herself has led mates on long sojourns through adjacent home ranges. Alone she might have been threatened by resident bears, but with the protection of a mate she was able to access potential food sources without conflict.

I've also had the chance to see bits and pieces of pre-courtship, courtship, and mating–and again Squirty has been my steadiest source of new information. In most years, her potential mates begin showing up at the clearing about two weeks before she comes into estrus. When a male visits, Squirty trees herself with her cubs; the male back-rubs on several of the trees in the clearing before going to the base of the one holding Squirty and her cubs. She leans down, showing interest, and the male might stay around for an hour or so before moving on, not returning until Squirty is ready to mate. In some years, Squirty was ready before the male returned, and so she would separate from her cubs and go on tour around her home range to advertise her condition. Either way, she has always ended up with a mate. Over the years I have followed her, Squirty has had eight litters of cubs and nine different mates.

Males generally don't get to breed until they are around eight years old. They are sexually mature at age two, but the large males suppress the younger and smaller males' attempts to breed, and in the end just the largest 10 to 15 percent of the males successfully mate with mature females. I have seen subadult males traveling with subadult females during the breeding season, but these young females are not of interest to the older males. Lacking a home range, subadult females are unable to put on enough fat before hibernating, and without enough fat they run out of milk and automatically come back into estrus the next breeding season.

The cubs then starve, or are abandoned when the mother's mate shows up–or the male may kill them. Because males have no vested interest in cubs who are not going to survive, or who are not theirs, they may kill yearlings who simply don't understand the rules. Having said that, not all male bears kill cubs. I have remote camera images showing a breeding male feeding side by side with his mate's yearling. Try as they might, scientists have found no evidence that infanticide is sexually strategic in bears–in other words, they don't kill young for access to the mother. That notion is flawed on a very basic level. Even if a male bear successfully killed healthy cubs in order to mate with a female, once the female came back into estrus, her condition would be advertised on the airwaves and there would be no guarantee that he would win the opportunity.

One recent June, I was fortunate enough to observe pre-courtship, courtship, and mating behavior at a very close distance. Squirty, SQ2, and SNLO were all coming to the clearing with their yearling cubs. The male first showed up on June 9 after I had placed a roadkilled deer at the clearing. He checked out both SQ2 and SNLO, and both treed with their cubs while he was there.

He returned to the clearing on June 14 and came directly up to SQ2. This time her cubs treed, and she only partially treed up a cherry tree near my truck. She gave him some soft "huh, huh, huhs" then went over to feed ten feet from him. He left to check out Squirty's scent in the spot where she had lain down to eat the previous night, then went over to SNLO, who had her cubs above her in the big pine. He circled the pine, and she extended her body downward to check him out. She then came out of the tree and charged. He quickly retreated and returned to SQ2, who took refuge from him next to my truck. He made a tight circle around her, paying no attention to me, and she huffed at him to back off. He retreated, but did a full back-rub on a nearby apple tree before leaving the clearing.

In the meantime Squirty arrived at the clearing without her cubs, a sign to me that she was ready to mate. She checked out the male's scent, then chased and treed SQ2 before coming over to me and gently taking her Oreo treat from my hand. At this point, all of the bears at the clearing had settled down to eat at their respective spots. The male now returned to SQ2 and greeted her with soft seductive chirps–"mmh, mmh." She stood on the birch log and huffed at him in response.

He then diverted his attention to Squirty, who was lying down eating, and displaced her from her spot by walking directly toward her. Lying down, Squirty would have left a urine spot on the ground, and the bear wanted to check it out, to determine whether she was in estrus. He sniffed the ground, opened his mouth, and used his tongue to lift scent from the urine spot. She moved off about thirty feet and looked back at him as if to say, "Hey Big Boy, you're coming with me." He took up pursuit and made the gulp vocalization as he followed her out of the clearing. I didn't see either of them until June 16. SQ2 got the message and started chasing her cubs to break up the family so she could find a mate.

On June 16, Squirty returned to the clearing with that Big Boy and he started following SQ2. SQ2 responded with a "huh, huh, huh" vocalization, two standing face-to-face encounters, and a swat toward his head. Then she left the clearing, with the male following close behind. Squirty was unconcerned and quietly finished her meal.

On the evening of the seventeenth, Big Boy came flying back into the clearing, doing a rapid pass around it before heading to the pond to take a few cool-off laps. He had found mostly cubs at the clearing, so he left as quickly as he'd arrived. The cubs all scattered on his arrival and quickly returned when he left. I called him "the male on steroids." He was extremely focused on mating with as many females as possible and oblivious to me. He never once false-charged or chomped at me, and when we did meet at a close distance, he simply made a short respectful arc around me.

On June 18, the big male was checking out Squirty's scent and followed her out of the clearing. Lots of activity followed over the next few days, including the arrival on the nineteenth of SN2, Moose's granddaughter, and another large male. They were feeding together–a definite mating behavior–and she was periodically keeping him at bay with the soft "huh, huh, huh." He was staying close to her at all times and doing full back-rubs on trees close to her, essentially setting up a territory around her. On the twentieth, Big Boy came back into the clearing and marked over the full back-rubs of SN2's mate. On the twenty-first, he followed SN2 into the clearing and later left with her.

Thanks to the tenacity of Big Boy, I had seen more, in just this seven-day stretch, of the courtship and mating process than ever before. It wasn't until the twenty-third, though–when SNLO became receptive–that I got to observe the mating itself. SNLO came into the clearing with Big Boy in close pursuit. She came up next to my truck, and he approached her with his soft seductive chirp. He was bold and held his head high as she lunged at him and swatted to back him down. He made no sound, but stood and took her aggression with confidence. She then held her head low and emitted the soft "huh, huh, huh" repeatedly. She would move away and he would pursue. They'd also face off from time to time with open mouths. This went on all around my truck for about forty minutes before he was able to grasp her with his forepaws. At about 135 pounds, SNLO was dwarfed by this robust 350-pound male. Once in his grasp, she was unable to break free. He bit her behind the neck and proceeded to mate.

Bears are spontaneous ovulators. The female, once stimulated, produces an egg precisely when mating occurs, which ensures conception. While mating, ovulation, and conception come together successfully in the wild, this has often failed in captivity. Zoos have gotten better at breeding bears, but they were long challenged by the fact that captive bears would mate but often fail to produce offspring. Based on what I've observed in the wild, the problem lies in the fact that, in captivity, the mate can be too well known to the female and thus lacks the ability to stimulate her.

I watched the interaction between SNLO and Big Boy from the safety of my truck at distances of ten and twenty feet. It was clear that the bears were experiencing a great deal of tension and excitement as they came together for the first time, and all of that was played out in aggression. He was confident because of his size and experience. She, who was hormonally ready, had suddenly come into close contact with this enormous and persistent stranger. I could sense her internal conflict between wanting to protect herself and the excitement of the moment. Her refusal and delay were clearly enhancing his desire. The final grasp and bite to the neck was the trigger for ovulation to occur.

The first grasp lasted only a few minutes. Then SNLO broke free and led Big Boy off into the woods. They were only gone two minutes when two subadult males came running into the clearing to feed on corn. In most of the courtship and mating that I have observed, there is an entourage of smaller males who follow the dominant males around. These two young males did not stay around to take advantage of the food; they moved on when Big Boy moved on, as all subadult followers seem to do. I call this "mentoring," as there is a lot a young male can learn from an older, successful one. On my trail cameras, I've often seen a large male pass by and then, five minutes to an hour later, a smaller male follow along the same trail. Olfaction makes this possible: Bears employ their powerful sense of smell to track each other to find food, meeting places, and mates–and, as we will learn later, to inflict punishment.

After about a twenty-minute absence, SNLO and Big Boy returned to the clearing. The two males ran out of the clearing after Big Boy false-charged then chased one of them. Big Boy continued to pursue SNLO, acting out an exaggerated stiff-legged walk and marking with urine on either side of her in response to the other males before mating with her again. This time it was a prolonged act with multiple bites to the neck and bodily manipulation. They were attached for forty-five minutes. Literally attached. Bears have bacula–like the penis bones found in dogs–that lock onto the female pelvis until mating is complete.

They weren't alone in the clearing. There were cubs in the trees, and SQ2LO arrived and walked within fifteen feet of the lovebirds to get a good look. When they finally broke up, SNLO urinated and Big Boy spent three minutes assessing the urine spot on the ground. He was much more relaxed, but visibly panting. He let her eat and finally settled down to eat himself before coming over to her with the gulp vocalization encouraging her to leave the clearing. Eventually, the two disappeared into the forest together.

A whole new aspect of the bear world was opened up to me that night, and I began to understand the importance of female aggression in mating, which, I suspect, helps stimulate the female into ovulating so that conception is guaranteed. Timing is important when two animals come together briefly during the breeding season.

Alliances and coalitions exist outside the breeding season as well–not just between males and females, or between females who develop reciprocal relationships with each other, but also between males and males. I have witnessed these interactions personally, and have caught them on tape and film many times. Once, I had placed a small amount of corn in front of a remote video camera. It was an amount that easily could have been eaten by one bear. The camera turned on as a single large male entered the frame and began to eat the corn. About a minute later, a second large male entered the frame, initially avoiding the first, then turning and approaching with an open mouth–a bear's way of asking to share a meal. The two stood and wrestled, then both settled in to share the balance of the corn. Whether alliances like these are temporary agreements or long-term friendships, they are certainly reciprocal relationships.

The more I witnessed bears interacting, the more I realized how complex their social world actually is. I soon learned that I, too, was part of that complex world. Early in the spring of 2006, I hung a bag of corn from a tree in the clearing, using a tensioned horizontal rope that I could raise and lower. The bag was out of a bear's reach, but I left some wooden boxes nearby that the bears could move to get the food, although I fully expected that some of the bears would figure out other ways of getting what they wanted. I hoped their behavior toward the food, and toward me, would shed some light on their relationship with me. I also wanted to duplicate a famous experiment that tested whether chimpanzees could figure out how to use tools.

Many readers will recall that Jane Goodall found that chimps use sticks to extract termites from their mounds in 1960. But long before this, in the 1920s, Wolfgang Köhler had hung food out of reach of captive chimpanzees; he then scattered boxes around to see if the chimps had the insight to move and stack the boxes to reach the food. He also gave them sticks that fit inside one another, something bears could not manipulate without opposing thumbs and fingers. The chimps were successful in attaining the food, both by stacking boxes and by assembling sticks and knocking down the food. Would the bears at the clearing simply move the box–in this case, their tool–and stand on it to get the food? I had no great expectations, though bears' penchant for creating a nuisance around human food sources suggested they might have some interesting ideas about how to get to the bag.

It turned out the bears had no trouble getting their paws on the corn. On her first try, Snowy sized up the height of the bag by standing on her hind legs, then bent at her knees and jumped up to snag her quarry. But she still couldn't reach it, so she devised a better plan. It took her less than two minutes to climb the tree and snap the rope, pulling back with her teeth and letting go to cause the bag to break and release the corn. On another day, her daughter SNLO climbed the tree and pulled herself paw-over-paw along the rope to grab the bag, then dropped to the ground. Would Squirty display the same sort of problem solving? I wondered.

I stood by with my video camera as Squirty approached the bag for the first time. She had already eaten, so she simply stood on her hind legs to check things out, then walked away. The next day, I hung a bag once again and waited with my camera. This time Squirty was hungry. She stood on her hind legs looking at the bag, then went to one of the boxes, rolling it in the wrong direction and biting it. She then came over to my truck to see if I was hiding food from her. After looking in all of the windows, she placed both paws over my mirror and started to rock the truck, all the while looking up at me. I knew where this could lead. On a previous occasion, I had left Squirty's corn on the seat of my truck with the windows up and gone for a short walk. Upon returning I found my truck missing both mirrors. Apparently, Squirty was telling me that I should have left the food where she could get it. So when Squirty started rocking my truck, I quickly yelled for her to stop.

Squirty then went back and started to mess with my remote camera. After a moment's pause, she proceeded back to me, giving me first a mock-symbolic bite in front of my knee, then the real thing, puncturing the skin on either side of my knee. I produced the bag of corn, and she reconciled, emitting a series of soft moans. Her message was clear: She was sorry, but the bite was necessary. I was less shocked by the bite than by the insight it offered. Though Squirty was obviously capable of getting the food herself, she was punishing me for making her job difficult. I finally understood that our relationship was a contract, one built on mutual trust and maintained with pain and pleasure. Two days later I repeated the experiment with my camera running, and was promptly false-charged by Squirty, apparently for breach of contract. We had a clear relationship, established through years of trust. And I was not holding up my end of things. I did not try this again with Squirty; getting bitten would not accomplish much. But I do intend to set up the experiment again when the remote camera technology improves and I can ensure capturing all of the results.

Before this experience, I told people that the difference between my relationship with a dog and that with a bear was that a dog would accept whatever I did. My relationship with the bears was different, I knew, but I couldn't articulate it clearly. Now I could.

Squirty and I had drawn up a social contract not with words, but with actions. It was a cooperative relationship, a tit-for-tat that was reinforced with the skillful application of reward and punishment, ranging from explosive aggression to passionate reconciliation. It made no difference that the contract was between a man and a bear. What was significant was that both parties were players, something I might not have recognized had I not been one of the players myself—with the scars to prove it. My mother–daughter relationship with Squirty had drawn me in to the heart of the bears' world, and proven key to my understanding of black bear behavior.

Chapter Four

The Language of Bears

Being adopted into Squirty's world has left me with many puzzles to solve. Like a stranger in a foreign land, I've had to piece together clues to find out that bears, often described as solitary, are anything but. As the daily life at Lambert clearing shows us, Squirty and her counterparts form matriarchal hierarchies of related females, have friendships and alliances, and routinely interact with unrelated bears. And it is by watching all those interactions, and also paying close attention to how bears interact with me, that I've begun to decode the language that supports their complex social structure.

What I have found is a rich mix of communication–both emotional and intentional.

Yet many of the books on animal communication that line my shelves claim that animals communicate only with emotion, not intention. In fact, most of science lumps animal communication under the banner of "motivational," which includes emotional communication and communication designed to *convey intention*. It steers clear, though, of the idea that animals communicate with full knowledge of the messages they are sending and the impact they will have on those who receive them. Consider how psychologist Robert Seyfarth and biologist Dorothy Cheney characterized animal communication in the February 2003 *Annual Review of Psychology*. "The inability of most animals to recognize the mental states of others distinguishes animal communication most clearly from human language," they write. "Whereas signalers may vocalize to change a listener's behavior, they do not call to inform others. Listeners acquire information from signalers who do not, in the human sense, intend to provide it."

These researchers are far from alone in believing that no form of intentional communication exists in the animal world. There is some debate, though. Scientists like Donald Griffin and Frans de Waal, for instance, have made recent strides in showing that animals show clear signs of carrying out complex thought processes–from demonstrating an ability to understand the future consequences of their actions to showing the ability to consciously "lie" when such trickery is to their advantage. Again, most people wouldn't question such cognitive skills. But skeptics remain.

To me, it has long seemed obvious that most intelligent animals communicate with both emotion and intention, in various degrees relative to their need, and my observations of bears have underscored that assumption. Because bears need to regularly cooperate and interact with non-relatives and strangers, they have evolved the sort of advanced cognitive features that likely sparked the evolution of human language. As our own society became more complex, with an increase in density of human populations, the distinction between emotional utterances and intentional, mechanically generated sounds may have slowly integrated into language as strangers needed to interact. That's a bold assumption that we'll explore later on. But for now, if you're wondering why social interaction makes a difference in the way an animal communicates, think of it this way: If you're going out and interacting with strangers each day, as bears often are, you've got to be able to communicate using more than just emotion. How else would bears communicate with strangers and manage their food sharing and open society?

So how do bears get their messages across? They use ear, eye, eyebrow, and facial expressions. They also use a wide array of emotional vocalizations, which they emit at different levels of intensity, such as when they are reacting to danger. And they rely on certain sounds (like their breathy "huh, huh, huh") and actions (like their stiff-legged walk) to convey intention rather than emotion, with different meanings depending on the context. Bears might use these signals to curtail aggression, show that they are backing down, intimidate other bears, or simply cause another bear to change its behavior. By combining emotional and intentional forms of communication, bears can build complex social interactions.

If all else fails, bears will act out what they want to communicate. I experienced this with the cubs I raised as well as with Squirty when she was an adult, and it has also been reported in zoos between bears and their keepers. Scientists believe that pantomiming represents the precursor to language, but whether nonhuman animals are capable of pantomime–and if so, which ones are–has also been the subject of debate. Despite many descriptions of great apes acting out messages, it wasn't until early 2013 that the first "official" study on pantomiming in orangutans was published in the *Royal Society Biology Letters*. Still, some bears seem to pantomime quite clearly.

Within our own language, we find the same blend of emotion and intention. Some might call this nuanced, strategic communication; and as far as I can see, we haven't cornered the market on it.

Bears, I have found, have two basic means of communicating: with scent (which includes both leaving it and interpreting it) and without scent (which includes vocalization, facial expressions, intentional expressions, body language, symbolic communication, pantomime, ear expressions, and many combinations of these non-olfactory cues). Let's explore scent first.

Everywhere a bear travels it leaves, unintentionally, a scent that can be identified and followed by any other bear, using an innate system of chemical analysis to differentiate between smells. This transparency allows bears to follow each other to surplus food sources and to detect and hunt down cheaters, or freeloaders, who break the rules. Squirty always knows exactly who has broken her rules; she doesn't need to guess or do detective work. She can track down and punish the offender as often as it takes to make herself understood.

There has been a bias in science that has dismissed the notion of any olfactory-oriented animal having the cognitive ability to behave in ways that might have led to human development. In his book *How the Mind Works*, Steven Pinker, the celebrated evolutionary biologist from Harvard, puts it this way:

> Most mammals hug the ground sniffing the rich chemical tracks and trails left behind by other living things. Anyone who has walked a frisky cocker spaniel as it explores the invisible phantasmagoria on a sidewalk knows that it lives in an olfactory world beyond our understanding. Here is an exaggerated way of stating the difference. Rather than living in a three-dimensional coordinate space hung with movable objects, standard mammals live in a two-dimensional flatland which they explore through a zero-dimensional peephole.

What I've learned on my journey with bears, though, has suggested that their world is not the two-dimensional flatland that Pinker describes but rather a four-dimensional one. When I began exploring that world, I knew bears' behavior was a system that had been finely tuned over millions of years of evolution, and I suspected that it could be described with the observation of enough of its pieces. And so when it came to understanding the many different ways they communicate via scent, I attempted to decode every scent-related behavior I observed, and to relate it to every other piece of the puzzle I was assembling on ursine life. I sorted every behavior I observed into broad groups, then into subgroups. I explored whether the marking I observed was intentional or unintentional: Was it done for a reason? Or was there a reason for transparency? Each time something new developed, I would test it against other observations I had made, searching for patterns and meaning in the relationships. That part was easy for me, as it was how my mind liked to function.

I've known for a long time that all animals leave scent trails when they travel, but not all animals are equally endowed in following scent trails. As humans we leave an airborne scent trail wherever we are and scent on whatever we touch, yet we are not equipped to follow those trails ourselves. A bloodhound can distinguish minute amounts of an individual human scent and follow it for miles. The scent animals and humans leave as they travel is unintentional; there is a transparency about it. Like the GPS systems in automobiles and cellular phones, scent records every activity of the individual.

In the human world, scent is of very little consequence, but in the world of the bear it is the fabric of their judicial system. Bears' ability to receive and analyze scent puts bloodhounds' to shame. I watched the cubs I raised sniff, lick, and huff (lifting scent off objects with the moist breath from their lungs) scent on every twig, leaf, or blade of grass that another bear's hair had come into contact with, leaving behind minute amounts of sweat, sebaceous oil, urine, semen, and hair or skin cells. They would follow the scent to see what the bear was up to and what it was feeding on.

An incident I observed in the summer of 2010 at the clearing shows just how bears use scent to discern events and track down other bears. Two, Squirty's granddaughter, had finished the small amount of corn I had given her and gone up the hill to take a nap. While she was gone, SQ2's daughter SQ2LO came over to clean up the residuals. It was not unusual for subordinate females to push the boundaries of those above them in the hierarchy. When she had finished, she wandered off into the forest. After a while, Two returned to find her residuals gone and the scent of the bear who ate them. Two tracked SQ2LO step by step into the forest and settled the score.

Bears, it seems, cannot hide from one another. There is no sneaking around, no traveling incognito. Even their long-range communication is transparent. Every hair on a bear's body has a combination of sweat and sebaceous glands. The sweat glands excrete aromatic compounds that respond to the bear's central nervous system and can signal alarm or indicate where a female is in her estrous cycle. These light and airborne molecules travel in streams and puffs in the wind–or they can settle and flood an area on still days, carrying information about the bear's identity, gender, genetic relatedness, and hierarchal status. Wherever a bear travels, it broadcasts this information onto the airwaves. Every bear knows not only when another bear has entered its space, but also who the bear is. And it can follow, approach, or avoid that interloper as the situation necessitates.

It also knows if a nearby bear senses danger, as bears emit two levels of alarm scent–one from the sweat glands that broadcasts a warning, and the other from the anal glands, released in moments of more extreme fear. Both are strong enough to be detected by humans. One day I was making my way up Lambert Ridge on snowshoes, following telemetry signals to get to Snowy's winter den. Now, I have been to many winter bear dens and have never been able to locate any of them by smell. In fact, even when we have sedated bears in their den to change their collars, weigh them, and take hair and scent samples, I have to put my nose right into their fur to smell their sweet hay-like scent. But this day would be different. Seventy-five yards from the den I got a strong blast of musky scent–the kind of alarm scent emitted via sweat glands. I was concerned and feared that something had disturbed Snowy and her cubs. When I got to the den entrance, the cause was obvious: Two coyotes had walked within six inches of it. It was a March afternoon and the snow was melting. The tracks, in near-perfect condition, told me the coyotes had passed in the previous fifteen to twenty minutes.

The role of sweat-gland alarm scent became especially apparent to me during my many walks through cornfields in search of bears. It is not unusual in a year of poor natural food to find as many as twenty bears in a forty-acre cornfield. The cornstalks are so dense that visibility is reduced to about ten feet. When one bear is scared by a predator, its scent spreads rapidly through the dense corn, alerting other bears. Cornfields are not the only places bears congregate, of course; the same warning system comes into play when they are gathered in berry patches, beech and oak stands, and other areas of concentrated food.

When bears are really scared, though, they release scent from their anal glands. These, like the anal glands of domestic dogs, put out an unpleasant and powerful odor. I have only experienced this smell from bears on two occasions, and both were when I was catching bears with a barrel trap in order to fit them with the telemetry or GPS collars that aid my research (and also distinguish the bears I work with from the general hunted population, helping to keep them alive). Most bears handle being caught with little effect. Some are even asleep when we get there, and others keep coming back for a free meal knowing they will be released unharmed. But that wasn't the case in 2001, when I set a trap at the clearing in the hope of catching Snowy. I use trail cameras to determine if the bear I would like to catch is in the area, but I can't be sure that it will be the only bear lured by the trap. When I arrived for Snowy, there were several other bears at or near the trap. Squirty, her mother, was sitting about thirty yards away, and a two-hundred-pound male was at the trap. It was then I first encountered the pungent alarm scent released from a bear's–in this case Snowy's–anal glands. I didn't get a collar on Snowy that day (I opted to release rather than sedating her with the company around), but I did pick up another clue about the role of alarm scent, suspecting that the strong anal-gland alarm scent had led the other bears to come to her assistance. That notion was reinforced the second time I encountered the anal-gland smell. Like the first time, I was checking a trap and found other bears there on arrival. It is possible that the alarm scent from the sweat glands acts as a warning of danger, while the excretion from the anal glands is an indication that an individual is in dire circumstances.

Like modern human language, olfactory communication in bears is a mix of transparency and deliberate communication. Just as bears leave an unintentional scent trail wherever they go, they also leave distinctly intentional olfactory communications. With sebaceous oil, sweat, urine, semen, and vaginal excretions, bears deposit scent marks to manage their world and reflect their position in society–not unlike human graffiti carved into trees or painted on subway trains, rocks, or bridges.

Their communications toolbox includes stiff-legged walks, full back-rubs, side-rubs, belly drags, and chin-rubs. Most people who walk through the woods in bear country have come across, at one time or another, what's called a bear tree. In my area, bear trees are usually red pines, and the telltale sign for those who happen upon them are the claw marks and bite marks etched into their bark. Both male and female bears will choose trees in their territory to mark in this way, and will visit them repeatedly to leave fresh marks. They also walk over saplings to leave their scent behind, or mark with scat or urine; and often they'll combine several approaches to convey the message "Kilroy was here."

If you separate the hairs on a bear's back, you will likely see a concentration of oils that have built up from the many sebaceous glands found in them. When a bear does a full back-rub on a tree, it is the scent these oils carry that will create its mark–and bears appear to go to great lengths to make sure as much of that oil ends up on the tree as possible. If the tree is large, the bear will squirm until it is sure that its back has full contact. If the tree is too small to support its weight, a bear will actually reach around with one paw and hold on to it to make sure all of its back makes contact with the sapling. Bears will even crawl beneath small trees to mark them. I have a series of pictures that show one female bear repeatedly crawling under a bent-over maple sapling only one inch in diameter, letting it ride over her back. It could be argued that she wasn't intentionally marking, but her repeat behavior suggests otherwise, as does the fact that on another evening I recorded a different female bear at the same sapling. She was standing and using both paws to hold the sapling against her neck and back as she flexed her knees, again making sure it had full contact.

Bears also do a lot of intentional marking when seeking mates. One of the ways a male bear will advertise his availability is to orally masturbate and then use his nose to deliberately smear his semen into the specialized scent-marking hairs on the sheath of his penis. He'll then mark by dragging his pelvis on the ground with his hind legs sprawled out behind him, propelling himself forward with his front legs. Semen, like urine, picks up the scent signature from the glands in the hair. By the end of the process, he has left a message on the ground: "I would like a mate."

As with unintentional scent trails, bears who come across these intentional marks can determine not just *who* left them, but also their gender, mood, relatedness, and social standing. It appears, too, that bears target their messages. While any passerby can decipher a mark, it is often left for a specific individual.

I know this because I have had messages left for me. For a few years, I was following Squirty with telemetry even when she was near the clearing and I always had a ziplock bag of corn on hand for her, which she'd eat by making a three-cornered tear in the bag and licking out its contents with the sticky saliva on her tongue. Sometimes if I had to leave before I reached her and knew she was in the vicinity, I'd leave the bag on a specific rock. But Squirty didn't rely on my knowing she was nearby, even if I had telemetry on my side. Instead, she would leave a fresh scat before I arrived, right where I park my truck. This happened often enough that I didn't think it was a coincidence, but I wasn't sure. Then one day I arrived and there was a fresh scat no bigger than a golf ball right on the rock where I would put the corn. Her message, and its target, were clear.

Squirty's medium in her message to me was scat. But bears use every tool in their toolbox to get their points across. Every year, after female bears go through the routine of chasing each other to establish their female hierarchies, the dominant female goes through a marking routine, often using everything from stiff-legged walks to full back-rubs and more. This lets the subordinate female know who it was that just dominated her, but it also conditions her to the individual scent of the dominant bear. At this point, chasing and other forms of standoff cease, and only the dominant bear's mark will be required to defend a territory, dominate a food source, enter a friendly female's home range, or let other bears know who gets to stay and share resources and who has to leave.

Bears use marks in as many conceivable contexts as there are contexts to conceive. When Squirty's first cubs, Snowy and Bert, were small, I would go in on her telemetry signal several times a week to visit with her. She had them at a big pasture; a nearby white pine served as a babysitting tree for the cubs. It was early in May, and Squirty was feeding on the tender shoots and leaves of a variety of plants and trees that were plentiful at the site. She had little need to move, and her cubs needed the time to develop their climbing skills and grow strong enough to travel. Squirty marked about a two-acre territory around the tree with sunken footprints, made as she marked and re-marked the perimeter of the territory with a stiff-legged walk, at the same time depositing tiny drops of urine from the specialized hairs coming out of her vulva. The sunken footprints, like blaze marks on trees, symbolized the boundary and carried a warning not to cross it, and the little drops of urine carried her social signature letting all comers know who they were dealing with. It was also apparent that she knew I might be followed. On my arrival each day she would pass by me and pound out a stiff-legged warning trail over my footprints.

Even a bear's saliva and breath carry personal information. When two bears who know each other come together after being apart, they greet each other with a nose-to-nose or mouth-to-mouth kiss or with open mouths. I have seen this between mothers and cubs, between siblings and friends, and between mates. Until I filmed this behavior at night, though, I didn't truly understand its role. The open-mouth behavior allows bears to exchange saliva or breath and identify each other at close distances, even in the darkness or when the wind is carrying their scent in another direction. Often it leads to a bipedal wrestle and the sharing of food or friendship.

It also took me some time to decipher why the cubs I raised mouthed all the new vegetation that they came across, often holding a delicate leaf in their mouths for just a few seconds, releasing it unharmed. They'd also stick their tongues out, using them to lift the scent from objects, then return them to their mouths with the tip falling in behind their incisors–a process that would be repeated a number of times until the bear seemed to get the information it was after. When I first observed this, many questions arose: Were they tasting the leaves? I ruled this out because the leaves were not damaged, just held gently in their mouths. What kind of response were they receiving? Did they have a library of smells in their genetic makeup?

I started recording their mouthing behavior and its outcome. When mouthing vegetation, or objects like frogs and mushrooms that I offered them, they seemed to be able to discriminate between what was edible and what wasn't, leading me to question whether that ability was aided by some special anatomical feature. I tested my dogs and myself for the same ability and found us lacking. I then decided to look for the feature that might be responsible for the behavior. That's when I asked New Hampshire Conservation Officer Tom Dakai for a roadkilled bear, did some crude dissection, and found a fleshy organ about the size of a jelly bean in a pocket in the vomer, a V-shaped bone that runs under the centerline of the nose and above the palate. Many mammals have a vomeronasal organ, also called a Jacobson's organ, that lets them lift scent with their saliva to detect pheromones and other chemical messengers. But as far as I could tell, such an organ had not been documented in bears.

So I continued the investigation. I studied a bear skull and found that the bone between the bottom of the vomer's pocket and the roof of the mouth was translucent, with an opening along the suture where the plates of bone come together. I assumed this must be for a sensory nerve used by the cubs when mouthing, but what the jelly-bean-shaped organ was, how it functioned, and how it was related to the nasal system or the vomeronasal organ found in other animals remained a mystery.

All this would take several years to sort out. I began my search at the Dana Biomedical Library at Dartmouth College, where illustrated books on the anatomy of the domestic dog showed how their vomeronasal organs consisted of two tubular structures that lay on either side of the vomer and passed through the palate just behind the incisors to reach a nipple-like papilla. I looked up everything I could find on the subject, and consulted a friend, antelope expert Richard Estes, who had documented the way antelope use their vomeronasal organ to process information when mating. But nowhere did I find anything that had any resemblance to what I had found with the bears. So I wrote to the author of one of the studies I had read–Charles Wysocki at the Monell Chemical Senses Center in Philadelphia–and was invited down to Monell to give one of their lunchtime lectures and hand-deliver samples of the organ. It was a bit intimidating presenting to a room with about sixty research scientists, and I heard some grumbling at the beginning about the fact I was not a PhD. By the end of the presentation, though, they seemed appeased and all stayed on another hour to watch footage of Squirty.

The samples I had of the organ were in formaldehyde, but only gross dissection was possible, so the sensory nerves coming into the organ were not intact. Nonetheless it was determined that the organ comprised fat, nerves, and blood vessels. After I arrived home, I received a letter from Wysocki stating that he had taken samples of the organ over to an anatomist at the University of Pennsylvania Veterinary School and received the determination that no such organ had been previously described in bears. Lacking any peers and hoping to protect my discovery, I named it the Kilham organ.

There were still more questions and the need to document the sensory nerves. By chance, I boiled a half-rotted bear head and found what I was looking for–a bundle of eight sensory nerves coming off the organ, running anteriorly up the vomer and through a small passage under the brain, and eventually spreading out across the roof of the throat. I realized based on the fact that the cubs had been mouthing leaves without damaging them that they must have been able to identify aromatic molecules with this organ, and the sensory nerves across the roof of the throat allowed them to identify airborne scent. And because of the location in the vomer, I suspected the Kilham organ was related to the vomeronasal system and found evidence of that in the boiled skull: There was tissue connecting the vomeronasal tubes in very small notches in the bone on either end of the pocket where the nerves passed.

It wasn't enough that I had seen these connections, I now needed to document them, which has proved much more difficult. I have worked with Jack Hoopes, a veterinary pathologist at the Dartmouth-Hitchcock Medical Center, dissecting and preparing histological slides to look microscopically at the tissue. We continue to try to get a picture of what I saw when I boiled that half-rotted head.

What I observed in the wild, though, began to make sense. I had watched bears opening and closing their mouths as they received long-distance airborne scent from other bears and tested it with their Kilham organ. They would false-charge or chomp if it was scent from a bear they didn't know or trust and would act indifferent or "smile" if it was a relative, a friend, or a mate. I have also learned that bears (except pandas) are the only known mammal with the Kilham organ, which I believe specializes in the identification of light or aromatic molecules. Interestingly, mammals that lack this organ all have long-range vocalizations. Dogs bark, wolves howl, pandas and chimpanzees call for mates, and so on. But bears are silent on the long-distance frontier. Contrary to popular belief in New England, they do not hoot.

As you can see, my method takes persistence, time, collaboration, reading, research–and it all begins with observation. It is unconventional, but for me it works. Slowly I am able to assemble the behavior, anatomy, physiology, and ecology of the bear into a model that makes sense. When trying to understand what I observe, I am comforted to know that the bear's behavior is an interrelated system that has been designed through the step-by-step process of evolution. I'm also aided by my earliest exposure to how designs evolve: my years designing guns and variations on guns that were marketed by Colt Firearms. It was then that I first applied myself to seeing patterns, piecing together systems, and pursuing leads.

I recall Austin Behlert, the pistolsmith I worked for in Union, New Jersey, saying on repeated occasions, "Boy, if someone could come up with that, they could make a million dollars!" Like me, he was a self-taught gunsmith and gun designer, and he always made that declaration when considering a problem that was very difficult to solve–but if solved would not only reap financial reward for the inventor but also open the door to many new developments. It was a simple statement, but buried within it was the understanding that each time a novel technology hits the market, there arises a proliferation of uses for that technology that continues until a new opportunity or problem is recognized and the competition to solve it starts again. That is the way of technological evolution.

Biological evolution shares many of these characteristics, with one life-form evolving from another over time. But it also differs from technological development in that problems are brought on by changes in environmental conditions, solutions often are the result of chance, and survival is the ultimate judge.

This was the new design code I had to understand. And one thing I wanted to know after learning why the cubs mouthed everything in sight was how a mother bear taught her cubs what to eat. In a mother's presence, I doubted that young bears learning the ropes would have to endure all the trial and error that my cubs did. The answer came to me by accident. While I was walking the cubs, I'd often get thirsty and go down on my hands and knees at a mountain stream to get a drink of water. When I'd do this the cubs would rush me and make an audible sniff, as if they were expecting something.

Again, it took a while for me to put everything together, but finally I thought maybe they were looking for chemical cues to what I was eating. So I did a little experiment. I got down on my hands and knees and started browsing on red clover, something I knew the cubs hadn't experienced and something I knew wouldn't poison me. Each of the cubs rushed to me, stuck its nose in my mouth, took a long sniff, and then immediately went out searching for and finally foraging on red clover.

This simple experiment told me three things. The cubs knew what I was eating was safe to eat because I was eating it. They knew how to identify it because of the organ on the roofs of their mouths. And most important, the speed at which they learned was evidence of imitation—something that marks the highest levels of intelligence in the animal world. After all, imitation is one of the ways humans learn, too: Our children do what we do. And it is one of the prime drivers of language acquisition in infants.

Within our own language, we find the same blend of emotion and intention. We all know that around the house and with our family members, the cries of emotion that are emitted by most animals would be enough for us to get along. Debbie can read my expressions and interpret my moods. She knows what I mean when I sigh or smile or what I might need if I called out in pain. And likewise, I can read her nonverbal cues. But when we go out into the real world, we have to get it right. We have to be careful in negotiating deals and social relationships. We also need to be able to deceive or bluff to ward off threats while at the same time revealing an honest representation of our emotional states in order to be accepted and trusted by those we interact with.

In the bears' world, honest emotional communication and the language of bluff and deceit remain relatively distinct and transparent. In our world the line sometimes blurs–though not always. When humans want to communicate with intention, they deliberately generate sounds and actions through physical means. Our language is an emotional utterance modified with the mechanical and deliberate manipulation of the larynx, lips, teeth, and tongue. With this combination we can deceive, bluff, and lie (all indicators of intentional communication). Yet our central nervous system still expresses the truth–such as when we try to stay calm, but our voice shakes, or when a polygraph reveals our emotions and exposes a lie.

Our language also depends on context. The meaning of our words depends on how and when they are used. The same holds true for the bears. Most of their vocalizations and actions are deliberate and mechanically generated. They chomp their teeth, huff by loudly drawing air in and out of their lungs, gulp by making a loud swallowing sound in their throats, take air into their lungs to make a reverberating sound in their chests, and among other things swat and false-charge. All these vocalizations and actions are delivered with different intensity (an indicator of emotion) at different times, and have different meanings based on context. They can also be used to deceive or bluff.

Even with our own advanced language skills, humans' most common mode of communication is emotional, and we send and receive these emotional communications on a subconscious level. To understand why, we likely need to understand just how critical it is for social animals like us to be able to convey states of emotion. Think about what happens when we gather as a group, for instance. We wouldn't feel comfortable doing it unless we were comfortable with one another's emotional states. We convey those states in large part by sending and receiving facial expressions. But while we might feel comfortable revealing our full range of emotions among family members and friends, doing so among strangers could be costly. It's no surprise, then, that when confronted with strangers, we use a neutral expression

I call "subway face." Anywhere in the world where strangers gather, this neutral expression can be seen.

Facial expressions can also reveal devious thoughts and deception in social interactions. If you go to a presidential rally, you will see Secret Service agents looking over the crowd. They are not looking for a gun; they are looking for a face that seems different from all of the other faces–one whose expression inadvertently broadcasts malicious intentions.

Some scientists have stated that bears are hard to read because they don't *have* facial expressions. This misconception is likely born from the fact that bears use their own form of "subway face" when they are unsure of a situation. I've come to recognize expressions in Squirty and the cubs that I've seen them use repeatedly, and have also seen other bears use. I've even found a great deal of similarity between human facial expressions and those of bears. There is no doubt a bear's expressions are harder to read, because of the facial differences, but they are there: Smiles are smiles and frowns are frowns. And I can always tell when I have overstayed my welcome when visiting Squirty. Her eyes twitch, and I heed her signal and depart.

There was a scene in the documentary *Grizzly Man*, about Timothy Treadwell's life and death among the brown bears in Alaska, that sent chills down my spine. In it, Treadwell was filming the same large male grizzly bear whom he had been filming for several weeks. His tripod was in the picture, his girlfriend was in front of the tripod, and the bear was sitting down about ten feet away. A close-up of the bear's face revealed that his ears were cocked back, a sign of annoyance, and his eyes were twitching, a sign that he was deliberating. I leaned over to Debbie and said, "If I had seen that, I would have packed my bags and left."

It was apparent from watching the film that Treadwell was not a student of behavior. He missed the telltale clues that might have saved his life. Sadly, it is not unusual for humans to approach wild animals with the same folk psychology they apply to their domestic animals. In this case, the casual approach to behavior proved fatal.

Treadwell had been pursuing the bear for several weeks, often getting pictures at closer than ten feet of a hungry bear searching for the last remnants of food. Bears communicate aggression or dominance and subordination with body movements. If a bear who is dominant wants to take food from a subordinate, it simply walks directly toward the subordinate, signaling its dominance. A subordinate bear can avoid being attacked by a dominant bear by walking in a wide arc, honoring the dominant bear. As humans, there are times when we may have no conscious awareness of what the movement of our bodies may be conveying to other humans, our pets, and especially to wild animals.

I suspect that the cocked ears and eye twitch were a signal to Timothy that he had overstayed his welcome. The grizzly likely interpreted Timothy's pursuit as aggression, and while the bear was tolerant for an extended period of time, he was close to having had enough. I believe Timothy mistook the confidence of a dominant male for habituation toward humans, and at some point crossed the line of provocation. It was not long after that scene took place that Timothy and his girlfriend were killed by the bear. If I ignore Squirty's eye twitch, her next action is a modified false charge, again based on the context of the situation, a message that it is time for me to leave. Early in the year, when Squirty's cubs are small, her tolerance for my intrusions lessens and her message for me to leave and–if I stick around too long–her false charge are much more intense. With this mix of intention and emotion, she and other bears are able to convey intent and an honest reflection of their emotion state.

An eye twitch is a signal that a bear is deliberating what to do about the problem at hand, and the problem at hand can be determined by the context of the situation. I witnessed eye twitching in a different context in early November 2002. Squirty was in the process of establishing and expanding her greater home range and had selected a den in her new territory–near Pout Pond in Lyme. It was a warm fall day, with about three inches of fresh melting snow on the ground, and I was making my way to her new site by following the steady beep from her collar.

When I arrived, I found Squirty in a den under a large, flat rock that was leaning against a short ledge. She emerged from the den to greet me, and I noticed right away that she had mud in her fur. The snow melting off the ledge was seeping down into the den–a condition that could be a problem in the spring when she had young cubs. I filmed her for about ten minutes as she stood on a stump and appeared to be quietly mulling things over. Her eyes were twitching. When she made her move, she went back to the den, checked the melting snow with her paw, climbed to the top of the ledge, and disappeared into the forest. I found her the next day two miles away in a rock den that she had previously used on top of Lambert Ridge.

Ear position, too, can clue an onlooker in to what a bear is experiencing or feeling. In fact, the way bears communicate with their ears is similar to the way horses and deer do. Bears have both functional and emotional ear movements. Some ear movements simply allow them to stay alert to sound. When they are eating, for instance, the very act of chewing creates noise, which compromises their ability to pick up other sounds around them. So they rotate their ears outward, with the openings opposite each other, for maximum coverage in picking up sounds. While investigating sounds or when alerted to scent, they rotate their ears forward in the direction of the sound or smell. When stalking another bear, their ears are also cocked forward.

Other ear positions, though, act out a kind of emotional sign language for bears. When they're approaching aggressively or attacking, their ears are pinned to the back of their neck. When irritated, their ears are half-cocked. A cautious but curious bear may have one ear back and one ear forward. Their ears may change position rapidly, reflecting rapidly changing moods. Some bears will position their ears nearly straight out on each side, looking like the *Star Wars* character Yoda. In fact, that's how the bear cub Yoda got her name. I haven't quite determined the meaning of Yoda ears yet; it might be just a friendly sign.

Bear body language can get even more expressive, though. Squirty has used and appears to make up gestures to get me to leave when she wants me to. One day when I was filming her in her den with her first litter of cubs, my video camera pointed to get the best view possible of the scene inside her two-by-three-foot excavation, she came out of the den and I filmed her as she simply clinched her paw into the snow. It was a clear sign of her displeasure. On several other occasions, she flapped her lips by blowing air through them–demonstrating a creative use of gestures to communicate with me. Creativity is one of the core aspects of language; in humans, it allows us to string words together in many ways to get our message across.

It may sound like Squirty is always asking me to leave, and it is true that my access into Squirty's world is an intrusion. It works only because the relationship is reciprocal. I pay for my intrusion with trust and food, her most limited resource. It is a pay-to-play relationship, and Squirty is constantly judging the fairness of the deal.

My reward, of course, with Squirty and the other bears I follow, comes in what I get to witness. On the communication front, some of my biggest surprises have come when watching bears use pantomime to get a point across. The use of pantomime–the art of using gestures, actions, and emotions to communicate without language–indicates intentional communication in that the sender must have a sense that the receiver has the ability to understand the information that is being communicated. The sender must be able to predict how the receiver might respond and have a sense that the receiver will understand.

One of the first times I saw a clear pantomime was when Little Boy passed me on a trail. It had been a long time since we had seen each other, and it's fair to say that both of us were very happy. I knelt down to greet him, and he suckled my ear. But I could tell he had a mate in the area, and so I didn't want my presence to disturb her, or for him to follow me and separate from her. When I tried to leave, Little Boy lay down in my path, pretending to be asleep. He wasn't trying to trick me into thinking he actually was asleep; he was acting out being asleep. And he was doing this based on prior knowledge of my behavior. He, Little Girl, and I had walked in the woods many times when they were cubs, and he knew that whenever we had, I would wait for him and Little Girl when they took a nap. His goal, it appeared, was to keep me from leaving. He did this on two different occasions weeks apart, but under similar circumstances. Each time he acted out similar scenarios.

In 2002 when Yoda had her first cubs, I went to visit her at her den in mid-March. I had brought her a small treat, mistakenly thinking that she would be hungry. She didn't want the food, but she did want my gloves. When I wouldn't let her have them, she acted out what she wanted by raking several beech leaves across eight feet of open snow to her den. The beech leaves were symbolic of her need for bedding. The lightbulb finally went off in my brain, and I checked her den—where I found that the winter snow had blown in and Yoda and her cubs were on ice. Yoda's action demonstrated her knowledge of my mind. Not only did she know I could understand her pantomime, she understood that I could deliver. I stripped off my fleece and donated it, along with my hat, to her cause. I returned the next day with a bale of hay, which she took from me and immediately started lining her den.

I am not the only observer who has documented this type of behavior in bears. Zookeeper Else Poulsen in her book *Smiling Bears* gives several accounts of a polar bear named Misty getting her attention with eye contact and acting out her need for fresh bedding. Wrote Poulsen, "I was beetling past the viewing windows, going someplace else, when Misty came running up. Her hurried manner and determined stare demanded my attention. I stopped and gave it to her. Keeping her eyes locked with mine, she backed up to the den entrance, raked some dirt, picked it up in her mouth, brought it over, and then dropped it in front of me." Initially, Poulsen didn't get the message. Then Misty tried again when Poulsen showed up next. "Again she hurried back to the window, locked in on me with her stare, backed up to the den entrance while still staring at me, raked up a small pile of straw, picked it up in her mouth, and brought it over to the window," she recalled. "This time she lifted her right paw off the ground and dropped the straw on top of it—some fell off, some stayed. Then it hit me. I had forgotten to give her a new bale of straw." Poulsen saw an opportunity for an experiment and held off giving Misty new straw until she asked for it. Sure enough, Misty did just that.

Bears also have a repertoire of vocal sounds to get their points across. Male bears use a loud gulping sound to get female bears to go with them during the mating season. Female bears use a similar gulp vocalization coupled with various body movements to guide their cubs safely through life. When a mother perceives danger, she will make a gulp vocalization to signal her cubs to group tightly and follow her. If she gulps walking toward a safety tree, she wants them to go up it. If she gulps at the base of a tree, she wants them to come down. And when she walks away and gulps, it is a signal for them to come down and follow. If there is danger, there will be an urgency in her gulp and in her actions; she may spin around gulping and run for the tree she wants her cubs to go up. It could be argued that this linkage between the gulp vocalization and body language constitutes an early form of syntax (the way words are put together to form phrases).

One summer night when I arrived at the clearing, I watched Squirty as she approached my truck. Her two young cubs were running wild, each going to the base of a separate tree. When Squirty spotted SQ2 at a fairly close distance, she gulped and turned at the same time. Both cubs immediately fell into tight formation behind her as she navigated them around the back side of the truck to avoid any conflict with SQ2. I thought back to the time I spent walking cubs in the wild and wished that I had that same ability.

I am not convinced that cubs are born knowing how to decipher the mix of gulp vocalizations and body language that their mothers use to guide them to safety. The "training" process I've observed mothers and cubs go through suggests that this particular knowledge is more the product of cultural transmission than of instinct. I spent a lot of time with Squirty and her first cubs, documenting their behavior, and watched as she false-charged her cubs to prevent them from following her and to send them back up the babysitting tree. The false charge was as convincing to the cubs as Squirty's false charges had been to me in her repeated attempts to modify my behavior. I had much greater difficulties in my own attempts to teach the cubs I raised how to respond in dangerous situations like road crossings or encounters with other people. But the one thing that I did learn was that cubs, like human children, do not come into this world trained.

It is likely that the basic bear vocalizations are instinctive, as they are evident in black bears throughout the country, but their use may be learned. Some vocalizations are clearly with the cubs from birth, though–something I learned from raising orphans and confirmed when observing them in the wild. When they are nursing, or suckling on my finger or another cub's ears, I've heard cubs sound out a rhythmic, droning sound of contentment–something like a deep, loud purr, and a sound they'll emit later, too, when happy or content. When falling asleep or during quiet times of joy, cubs will make a sound like pigeon's coo. Their distress call sounds like "Baa WoOow, Baa WoOow," and if they are in emotional distress they'll emit an undulating moan. When the cubs express recognition, they let out a series of soft moans. I've also heard cubs use different types of moans to broadcast hunger, nervousness, and irritation. When very angry, they roar. They gulp when nervous; they vocalize an "eh-eh" when wrestling to limit biting or to call an end to the ruckus. And they'll sound out a "mew-mew" to solicit the company of a nearby bear.

Adult black bears I've observed also use a guttural reverberation of sound in their chests to voice displeasure. Again, the intensity of the sound conveys the level of emotion, and the context of the situation rounds out the meaning. Like the gulp vocalization, the chest reverberation is deliberate and is mechanically generated with the chest and throat. One evening at the clearing I filmed the perfect example of one as SQ2 went through her routine of chasing and treeing her yearling cubs. It was September and time for her son BB to go out on his own. SQ2 was ratcheting up the pressure to get him to leave the greater home range. She had put him up the cherry tree by my truck and then chased and treed SQ2LO, his sister, up a nearby pine. He tried to come down, but the sound of his claws on the cherry bark attracted his mother's attention. She returned and chased him up the tree then grabbed on to his hind leg and dropped from the tree, ripping him out of the tree with her as they headed for the ground.

At the base of the tree, the two faced off in a bipedal stance. SQ2 stood on one side of the tree, and BB on the other. Both bears' heads were pointed down, and they were loudly vocalizing with the guttural reverberation. From the context, it was clear she wanted him to leave the area and he wanted her to back off. The deliberate nature of the confrontation became apparent when she approached me seconds later with a completely calm demeanor and asked for an Oreo. Realizing the opportunity, I kept the camera rolling, reached into my truck for an Oreo, and hand-fed it to her. I was stunned how quickly she could go from a highly aggressive situation to complete calm, as if she had just been doing her job as a mother. She wasn't emotionally charged as I would expect. I am not sure that we, as humans, could compartmentalize and control our emotions as cleanly as SQ2 did.

I, too, have experienced the bear version of "catching hell." One March, when most of the snow had melted on the hill where Squirty had her den, I packed extra camera gear in the hope of filming her while she was still inside. She was already out in front of her den when I approached, and she waited as I made my way up through the rockfall to reach her. When I was about twenty feet away, she came to greet me with a soft "mmm, mmm, mmm" appeasement call. I quickly took off my backpack and extracted the bag of white grapes I'd brought as a treat for her. Apparently grapes weren't what she had in mind, though, and she made a move for my whole pack. I tried to keep it from her: it didn't contain any more food, but it did have about four thousand dollars' worth of camera gear. The fact that I was leaning against a tree to maintain my position on a very steep incline gave her a distinct advantage. She grabbed the straps of the pack and pulled. I held on and shook the pack in the hope the cameras would come out. My telemetry receiver fell to the snow, but my cameras stayed put. Squirty threw a more aggressive "huh, huh, huh" into the mix, and I let go of the pack. I knew what that meant, and I wasn't about to escalate the encounter.

Squirty made her way back to the den with the pack, and I winced every time I heard the metallic ring of my cameras as the pack banged against the rocks. It was a cloudless spring day, and the temperature was about sixty degrees. There were still patches of snow around, and as it melted it dripped loudly on the ledge above me. The shadow of a large bird passed across the stone. I looked up to see a turkey vulture glide across the thermals created by the rising heat along the ledges. Squirty, sitting comfortably in front of her rock fortress, was working my pack over with her paws and sniffing it thoroughly with her nose.

This was a scene that could have played out in an airport; I wondered who let her know about increased vigilance and homeland security. When she was done with the pack, she got comfortable and I got my first look at three little cubs. They were obscured by the pack, but crawled up into her fur when she lay back. I spent about two hours within ten feet of Squirty as she sunbathed and napped with her cubs. If I moved, rustled leaves, or snapped a twig, I was reminded by Squirty with a squared-off lip and a chomping of her jaws that I was a guest and I could be asked to leave. It was also a reminder that I wasn't going to be able to just climb over to her and retrieve my pack.

At the end of two hours, Squirty got up and gave me an opportunity to film just the cubs. I heard soft moans coming from her so I didn't give leaving much thought until I swung the camera toward her when she was about three feet away. She took offense and aggressively (with a "huh, huh, huh") asked me to leave. As always, Squirty was in charge. And not only was she in charge, but she had secured some very expensive toys for her cubs.

I schemed overnight on a strategy to retrieve my cameras. I decided to return the next day with a bribe and hopefully catch Squirty away from the den. The plan succeeded: She was at the base of an ash tree when I arrived, giving her cubs their first climbing lesson. The cubs would not climb far up the tree and were tentative in their motions. When they got scared, they bawled, and Squirty stood at the base of the tree with her arms stretched out toward them to comfort them, helping them to the ground when they reached her. While she was occupied with the cubs, I gathered my cameras, which were lining Squirty's bed at the den entrance. Aside from a tooth mark or two, they weren't much worse for wear. The eyepiece of one of my video cameras was chewed off, but the camera still functioned.

The whole encounter provides a great lesson in bear communication. Remember, when one bear walks directly toward another it is a sign of dominance or aggression. So a bear needs a modifier to signal when the approach is non-confrontational. That's why when Squirty came to greet me once I reached her den she used the soft "mmm, mmm, mmm." She was signaling that her intentions were friendly. Bears use a similar "mmm, mmm, mmm" vocalization as a plea for reconciliation after they issue what I call moralistic punishment–in other words, punishment meant to enforce codes of conduct but not to permanently alienate the recipient.

It is interesting to note that "mmm, mmm, mmm" and "huh, huh, huh" are very similar vocalizations, except one is made with the mouth open and the other with the mouth shut. The meanings are opposite, with the guttural "huh, huh, huh" being negative, the human equivalent of a chewing-out, and the soft "mmm, mmm, mmm" being positive–either nonthreatening or conciliatory.

You'll notice that I've described some of the black bear's communication behavior as symbolic. But symbolism is one of the traits of human communication, a trait thought confined to the human mind. So how can it be attributed to bears?

Scientists can only speculate how traits like symbolism might have evolved. The same is true for me. Yet the more I watch bears, the more I question assumptions about what is uniquely human. Do bears and humans share some fundamental expressions of traits–traits that we regard as purely human? I believe we do, and that's because I have seen methods of communication thought to be uniquely human being played out by bears. I've also seen them use every means possible to communicate with other bears: emotional or honest communication, deliberate and intentional communication, transparency, and early forms of syntax. And yes, I've also seen what I consider to be symbolism and creativity.

All of this leads me to ask: Are we seeing in bears the fundamental elements of the platforms necessary for language to evolve? Could bear communications hold clues to the earliest forms of our own language? I believe the answers to these questions might lie in the similarities between the forces that shape bear society and shaped early human society.

Yoda reaches up to greet Ben, who raised her when she was orphaned and reintroduced her to the wild. BILL GREENE, *BOSTON GLOBE*

Nursing orphaned cubs is a round-the-clock affair. DEBRA KILHAM

When the cubs reach about four months, they move into a wooden enclosure. LAUREN GESSWEIN

Phoebe, on a cub walk, stops to let Big Girl check out some saplings.

Cubs quickly learn to tree when startled.

Yoda greets Ben on a wooded path, her cubs trailing behind. BILL GREENE, *BOSTON GLOBE*

Josie, one of Squirty's granddaughters.

Squirty and Josie, the cub she adopted from SQ2, who was struggling to care for three cubs.

Ben checks out a den Yoda used for two years—when her cubs were newborn, and when they were yearlings. BILL GREENE, *BOSTON GLOBE*

So many orphaned cubs arrived in 2012 that when winter arrived they just kept each other active and refused to hibernate.

The bears that Ben raised still treat him like a bear, and wrestling is a common form of play. BILL GREENE, *BOSTON GLOBE*

SNLO keeps Big Boy (right) at bay until she is ready to mate.

SNLO and Big Boy (right) in their precourtship phase during mating season.

Males and females can travel together for many days before mating.

The two courting bears give each other an open-mouth greeting.

SNLO and Big Boy mate. While mating occurs in spring, the fertilized eggs won't actually implant in the female's uterus until denning season arrives—and only if she has gained enough weight.

In an adaption of the classic mirror test, Ben showed that bears have the capacity to recognize themselves.

Ben tracks radio-collared bears in his study area. BILL GREENE, *BOSTON GLOBE*

Some of the bears, like Squirty, have GPS collars. They weigh about two pounds.

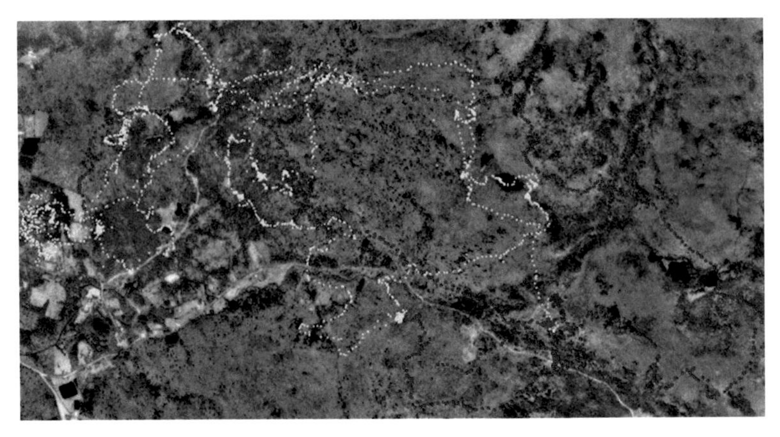

GPS can provide detailed data on a bear's movements. Here, GPS tracks Squirty while she traveled 14.2 miles in one day, using a big male as cover.

Ben comforts a sedated cub, just brought to him by wildlife officials after he was hit by a car. BILL GREENE, *BOSTON GLOBE*

When orphaned or injured bears are yearlings and ready to return to the wild, they are trapped, then sedated and relocated.
BILL GREENE, *BOSTON GLOBE*

Debbie Kilham keeps a cub warm while her sedated mother is checked during a den visit.

In winter, Ben and Andy Timmins (right), New Hampshire Fish and Game bear project leader, check Ben's study bears in their dens. Here Andy and Will Staats remove a sedated bear from its den to record its weight and other vital signs.

Mother bears scoot young cubs up a babysitting tree when they need to feed, or when danger lurks.

An adult male bear track, shown beside a pocket comb for size reference.

Squirty, shown here as a cub, came to Ben as an infant and has now spent sixteen years in the wild–giving Ben a unique window into the bear's world. Watching her mouth leaves led Ben to discover the Kilham organ.

Cubs are born in their mother's winter den and generally stay with her for the next eighteen months.

The bonds between Ben and the bears he played mother to remain strong, and they often greet him bear-style by sniffing his breath.
BILL GREENE, *BOSTON GLOBE*

Chapter Five

The Social Code of Bears:

Justice and Punishment

There has been a common thread that has run through my relationships with all the bears I have worked with: their constant effort to mold my behavior to benefit them the most. There has been a continuity of gestures, message bites, swats, false charges, mock bites, scent marks, and vocalizations that ranged in intensity from the gentle sounds of appeasement vocalizations to explosive attacks all aimed at getting me to conform to their wants and needs. They've punished me when I've stepped out of line–often, apparently–and they've been conciliatory when I've corrected my behavior, as a means of repairing any fallout from the punishment.

On the other side of the equation, I've been manipulating the bears for my own selfish needs. In order to observe them up close and learn from them, I've had to gain their permission: to infringe on their security by bringing in strangers (like other biologists and film crews), to outfit them with radio collars, and to generally take up the time that they would otherwise put to use simply surviving. I paid for these things with food, trust, and transparency.

These relationships have been built on mutual trust. Because I raised the cubs, they knew me better than I knew them–just as with human children, who often know exactly how to push their parents' buttons. Between the cubs and me, an unwritten and unspoken contract had developed, enforced with punishment and reconciliation. I recognized this type of contract: I had it at home with my wife, Debbie. From the time we met, we unwittingly began defining and enforcing the terms of an instinctual contract, the blueprint for a cooperative relationship. Ours involved, among other things, compromise, fidelity, a division of duties, and the preservation of our individual identities. This contract involved words, yet you could replace the words with emotional utterances and our relationship would still function, using nonverbal mechanisms of sticks and carrots, pain and pleasure, anger and reconciliation. Despite humans' penchant for legal contracts, we still to a large degree manage our relationships the truly old-fashioned way, the way they were managed before humans spoke their first words.

We all experience this in our everyday lives and constantly calculate the value of our relationships in terms of costs, benefits, and fairness. Think of high-profile philanderers like Bill Clinton, John Edwards, or Tiger Woods and the decisions their spouses must make to stay or go. No two people, even in a marriage, know each other completely–and words aren't adequate to maintain a contract because, as we all know, people lie. My relationship with Debbie is managed like all human relationships: with expectations of certain behavior and actions. When my actions don't meet her expectations, her response is an emotional one that corrects me. It is sometimes harsh, because it can be–because in a close relationship of family or friends, outbursts can be atoned for and disputes can be reconciled. That's why we judge a friend more harshly than we would a stranger. It is much harder to reconcile with a stranger, and you never know when that stranger may be in position to help you;

it's safer to avoid alienating him or her. Despite all of our technological advances, we still manage social contracts in much the same way that the bears do.

First-time mothers are trained by their babies about infant care with alternating cycles of pain and pleasure. Babies cry when they are tired, they're hungry, or their diapers are soiled; they seduce their parents with smiles and pleasant interactions when things are going their way. This goes on through childhood and adolescence, but the methods change. The cubs we raised trained us as well; they routinely modified our behavior with punishment and good behavior. In contrast, the cats and dogs we have had as pets have been docile and tolerant of any and all of the mistakes we might have made raising them. Why is there a difference?

The answer lies in the type of social behavior that has evolved with each of the species. Cats are solitary animals who will fend for themselves and their offspring; dogs are social, living in packs with fixed territories, with their lives and resources controlled by a dominant alpha pair. Cats didn't need to evolve complex behavior to interact with others of their own kind, as much of their lives would be spent by themselves. Highly social animals like dogs and chimpanzees who live in small groups evolved to be tolerant in subordinate roles. The fact that young bears behave more like young humans suggests common influences upon their social behavior.

In fact, a bear's social framework may provide more clues to our social evolution than that of close relatives like chimpanzees, gorillas, orangutans, and bonobos. To understand why, we need to understand the bear's approach to morality (the sense of right and wrong), judgment (the determination of right and wrong), and punishment (the enforcement of right and wrong). It is argued that, at least in the human realm, these three things evolved in an altruistic social environment in order to prevent purely selfish behavior from dominating. That makes sense when you consider that we need to cooperate as much as or more than we need to compete to get along in our wider society. That is also true for bears, and the primitive social behavior I have observed in them holds, I believe, important clues to the evolution of human cooperative behavior, altruism, and moral behavior.

My relationship with Squirty and my understanding of how she manages her social relationships has left me with a much better understanding of my own world and the many layers of social complexity that I deal with on a daily basis. She has demonstrated the ability to make a contract with me and hold me to it; to selectively manage her relationships with her relatives and share surplus food with unrelated bears; and to manage her sharing relationships with judgment and punishment. She goes to great lengths to make sure she has primary access to food within her home range, but her ability to share also ensures that she had access to surplus food in a stranger's home range in a time of shortage. Her actions benefit her as an individual, but they also help her kin and strangers in time of need.

While Squirty's actions are altruistic and good, they come at a price. The price is her judgment about how to enforce her rules. These rules are not social standards shared by the community of bears; they are Squirty's and Squirty's alone. She makes them and she enforces them. This means that every other bear has this ability, and the ability of any bear to exist is based on its relationships with the bears it interacts with. It also means that a social code of cooperation that favors relatives and includes strangers is hammered out between individuals. These relationships are forged with each individual's expectations and willingness to enforce them. Enforcement is easier among relatives because of the trust fostered in a family environment. A full range of emotion can be displayed, and received in the context that it is meant. If the punishment is too harsh, it can be easily corrected with reconciliation. Punishment and reconciliation become the tools of managing cooperative relationships. Aggression toward a stranger would need to be measured, as the response would be less predictable and reconciliation more difficult.

This level of flexibility makes for a wide range of social behavior in bears: When their densities are very low they can be mistaken as solitary, but when their densities increase or when they congregate at a concentrated food source, hierarchies emerge as social interactions multiply. Increased densities lead to an increase in sharing and in the need to manage fairness and to punish those who are perceived to cheat.

The word *perceived* is crucial, because as humans we know that each of us can and does perceive things differently, and if we all acted on our perceptions the world would be in chaos. Like Squirty, we are all born with the need to survive and pass on our genes. And also, like Squirty, we were born to perceive, judge, and act on the judgments about the individuals with whom we cooperate. It would be nice if a precise accounting of costs and benefits were required to stir our moral compass, but it is not. Fairness is about what we perceive as fair, not necessarily what is fair. Our brain, like Squirty's, receives information and draws conclusions from that information. In Squirty's world, the information received comes only in the form of firsthand experiences and observations of others' actions. She acts on this information based on her expectations of another individual's behavior, which is based on a contract of cooperation–in other words, a social contract. Without one, there can be no rules.

For example, Squirty's daughters can share her home range as long as they obey Squirty's rules. If Squirty perceives a breach of contract, she will exact an appropriate punishment. Each bear has its own moral sense; it is both judge and jury. Rules become norms when environmental conditions cause a group of animals to come up with the same set of rules. This works in Squirty's world because each bear has social status and cooperates with a limited number of individuals, somewhere between five and twenty-five in the bears I have observed. The same idea still functions in our world: Each of us has our own rules, and norms change with changes in the cultural environment.

It is not hard to imagine how a system of rules would work without language. One of the most difficult things for any of us to grasp growing up is that there is a complex system of both written and unwritten rules that we must either follow or choose to break. Not only are our rules complex, but they change with time and cultural advances. It is possible that the length of the juvenile period in animals and people is directly related to the complexity of these rules. I know in my own life I discovered the existence of many of these rules by breaking them and being quickly reprimanded by those around me; it is no different in my relationship with Squirty.

As we have seen, a bear leaves scent everywhere it goes, intentionally and unintentionally, and that scent is what leads a bear to identify and track down those who break its rules. When Squirty's moral compass is stirred, it is not because she *thinks* she knows who broke her rules, she *knows* who broke her rules. There is absolutely nothing comparable to this level of accuracy in either human perception or our judicial system.

The bears have also a system of punishment that is commensurate with the infraction. I know as I have been on the receiving end of it many times and have the scars to prove it. On a number of occasions, Squirty has damaged my property as a means of punishing me. One night I chose not to go to Lambert clearing because of heavy rain, thinking that the weather would prevent Squirty from showing up expecting her food. I was wrong. When I arrived the next evening, I found the four-by-eight-foot mirror that I had erected in the clearing to conduct self-recognition experiments tipped over. It had been bolted to a section of building staging, and that wooden frame had been all bitten up. Another time I arrived at the clearing, checked for Squirty's signal with my telemetry receiver, and got no indication that she was around. I decided to go for a short walk while I waited to see if she would show up. When I returned, I found both of the rearview mirrors on my truck broken off.

As a human working with bears, I have also received more than my share of mock bites, message bites, and canines sunk into my legs–all sentences carried out by individual bears for my violations of their individual rules. I have never observed or received any aggressive action from a bear that lacked a clear motive. Being on the receiving end of punishing bites is less frequent for the bears. Much like predator–prey relationships where the evolutionary process prevents advantage from being too severe, subordinate bears are frequently smaller and more agile than their dominants. They can run faster, climb faster, and get farther out on a limb to avoid punishment. Bears have also developed a finely tuned communication system in which the subordinate individuals can read the intentions and emotions of their dominants and avoid their wrath.

The bites I have received were largely due to the fact that the bears who bit me treat me like I am a bear. Squirty, for instance, treats me the same way she treats her other family members–as do the other bears I've raised as orphans. Being treated like a bear has its rewards: For one, I have a fascinating window into the bears' social code. But it also has a clear disadvantage: Unlike the "real" bears, I am not fast and agile enough to get away when getting scolded. Whenever Squirty delivers a punishing bite or maneuver, she makes up for her episodes of moralistic aggression with reconciliation, cooperation, and kindness. She has even risked her own safety to defend mine.

This kind of treatment is far from what I have experienced with Squirty's offspring or other wild bears, all of whom have been too intimidated by my presence to apply their bear-to-bear judgments and punishments to me. However, it is possible that the bear's code of punishment may play into some bear–human encounters that end in tragedy. In the bear world the maximum sentence, as in the human judicial system, is the death penalty. It is used sparingly, but it is used, and humans have possibly suffered severe injuries or death when they unwittingly break bears' rules. My advice to those who wish to recreate in bear country is that they spend some time learning how bears communicate and behave so they can be proactive in de-escalating bear–human interactions to prevent the kind of human-and-stressed-bear situations that may lead to the use of bear spray or firearms and possible human injury.

In our world things are much more complicated, and perceptions of fairness are a lot less exact. Today, when people think about a contract, we think about a legal document that holds two individuals or entities to an agreement that has been negotiated with the use of language. These contracts are rigid and precise and allow for intervention by a third party, a judge, if one of the parties fails to live up to the terms. While our judicial system and advanced means of accounting allow us to trade goods and services without stirring our moral compass, we still make and resolve social contracts based on unwritten or spoken expectations. Actions still dictate compliance. When one person's actions defy another's expectations, raw emotional outbursts still define the terms of unwritten or spoken social contracts or agreements.

On a daily and often an hourly basis, we calculate the terms of the social contracts of those we interact with. From cradle to grave, we negotiate our position in life. Often, we don't do this with intent; we just do it. Despite our refinements, there is a quite a lot of the behavior that drives Squirty in all of us.

Human success, I believe, like bear social behavior, grew out of our ability to access and share surplus resources in times of need outside conventional home ranges. This ability is still the basis of our success as we trade for goods from all over the world. It seems logical that as tool use increased, the quality of our foods improved, and our populations grew, so did the number of individuals that each human cooperated with. As a result, social complexity and need for social control would have grown as well. With this came the need for and ease of forming coalitions. Coalitions form easily among cooperators as they all benefit from sharing, and they can and have formed among family and friends without the benefit of formal language.

It also seems logical that the cost of an increase in cooperators was an equivalent increase in individuals who perceived inequities and acted upon their perceptions. This increasing social complexity would have been adequate evolutionary pressure for better communication, social order, and a corresponding increase in brain size.

Consider for a moment that it took five hundred million years to develop the kind of social behavior found in our pre-hominin relatives before increasing technology allowed us to develop sustained surpluses. The density of their populations was limited by their access to resources; social behavior developed to maintain control and access to a limited amount of food. Each additional individual, at this early time, made a dramatic increase to the social complexity and created an evolutionary pressure for change. With better means to communicate, we would have been able to form larger and larger coalitions. Unfortunately, regardless of size, coalitions wield much more power than individuals, but, like an individual, they have the ability to perceive, judge, and take action. Within these coalitions there can be as many opinions as there are people, and the success of any opinion can be tied to the ability of any individual to promote it. The opinions or actions that succeed are the ones that have the broadest appeal among coalition members.

Coalitions can and have formed for many different reasons. Each independent country in today's world is a coalition of cooperators. Within this greater population of cooperators, smaller coalitions of people with similar interests have formed to compete for advantage with coalitions of differing interests. To make things more confusing, members often belong to many different coalitions that sometimes if not often compete with one another. This is true of simple friendships as well as special-interest groups that lobby Congress, fraternal clubs, scientific associations, or any other group of unrelated people who assemble for mutual benefit.

With improved means of communication, broad public support for an idea or ideology could sweep through a population almost instantaneously. Just like a flock of starlings that lift off together and fly in unison, a population of independent thinkers can suddenly think as one. Sadly, as history has shown, whole groups of people often don't have any better judgment than an individual. Collectively we have made as many bad decisions as good ones.

In Squirty's world, perceptions, judgments, and action take place on a limited landscape of firsthand information. In my world, perception, judgment, and action can take place with information that is first-, second-, third-, or more hand. Our individual moral compasses can be aroused very easily and often are, based on information that may be erroneous, incomplete, and biased. There can be many levels of ethical and moral behavior that are influenced by cultural, economic, and religious differences. To make matters worse, we promote what we believe and tend to believe what we hear. We join with others and rally for our joint beliefs. There are always two sides to these judgments: them and us.

The simple and limited results of sharing, judgment, and punishment in Squirty's world have given me great clarity in the ongoing discussion of nature and nurture. The fact that judgment and punishment are nature-based makes it much easier to predict, understand, and forgive my adversaries. It doesn't make being attacked any less painful at the time, but it does help to know that those who pass judgment often don't know why they do what they do or that there are moral consequences that they may never perceive for their actions. Among individuals, it is impossible to comprehend all of the differences in perception and judgment that may lead to an overheated moral compass. Might this raw emotional surge have developed over millions of years as a mechanism to prevent cheating, to promote fairness, and to ensure cooperation when food was shared?

Nature, it appears, was responsible for both our good and our bad behavior, and the job of nurture, our culture, was to promote the good and control the bad.

One day when I saw Squirty crossing the road in front of my truck near the Old Beal Cemetery, she was clearly agitated, looking back over her shoulder with her ears cocked back–behavior that typically follows an encounter with another bear. I drove up the road and parked at the next pullout. I had been experimenting with my homemade GPS collar, and I needed to fit her with a new one. As I followed her telemetry signal, my approach was loud and obvious. At the appropriate distance I called out to her. She was slow to respond, but finally I heard movement and saw her coming through the thick vegetative cover. She was still nervous and was high-heading scent that was arriving on gentle bursts of wind. I gave her a sweet treat, her price for tolerating the collar change, and successfully got her old one off and her new one on. I then gave her a corn treat, her price for tolerating me. What happened next brought me one step closer to my understanding of black bear aggression.

I sat beside her with my back against a large white pine tree. Without giving it a thought, I did something I had done a hundred times before: I reached out to touch her as she was eating. Only this time was different. She was still wary of the immediate presence of the bear she had just encountered. When a bear–or any animal, for that matter–eats, its concentration is diverted and its senses are compromised. Squirty's primary concern was that other bear. When I reached out and touched her, she exploded on me. I recoiled in the opposite direction and instinctively raised my arm in a defensive position, and it landed right in Squirty's mouth. She hesitated, then–as if to punish–she increased the pressure, sinking a single canine into my forearm.

The explosive moment passed quickly and Squirty retreated to her corn, but not before expressing apparent remorse and reconciliation for her actions with a series of soft appeasement moans. The explosive nature of the attack for something I had done many times before, coupled with the presence of the other bear in the immediate area, gave me clear insight into why situations like these can be so dangerous. With no clear and present danger, Squirty would expect me to reach out and touch her, an act of recognition. When there was a clear and present danger, though, it dominated her mind. She was prepared to respond with explosive force should her adversary attack while she was compromised. It is likely my touching her released her pre-programmed attack. The bite to my forearm was punishment for distracting her in a tenuous situation, and the moans of reconciliation were an indication of our close relationship.

This sheds light on why one of the most dangerous things to do around a brown bear, especially, is interrupt it while it feeds on a carcass. The feeding bear would expect other bears in the vicinity to detect the carcass's smell and, based on the high value of animal protein in the bear's diet, pursue it. So while eating, the bear remains ready to lash out at a potential competitor. A human who disturbs such a feeding bear would find it difficult to survive one of these assaults.

Bear-to-bear, the results would likely be different. As is often the case in animals that are well equipped to do each other great harm, a ritualized display or–as I like to say–a well-choreographed fight limits the amount of damage that can be inflicted on each of the participants. There was a time when humans could solve emotionally charged disputes with well-choreographed displays as well, but that all ended, as Konrad Lorenz points out in his book *On Aggression*, when man picked up the first rock to use as a weapon.

Choreographed aggression remains the preferred form of punishment among bears, though, and they begin to hone their skills at righting wrongs early on. The earliest scruff I witnessed was among my cubs, when one cub sensed that another was getting more than a fair share of toys. With an angry roar and swept-back ears, the offended cub tore into the other and a well-focused brawl erupted. All of their strikes were aimed at the head, neck, and shoulders–all areas that either had deep muscle or only shallow flesh over bone, limiting the injuries they could inflict.

Cubs also encounter aggression in the form of false charges from their mothers–disciplinary actions that keep their behavior manageable. The need for discipline in the natural world is very real; a cub who wanders off or up to the wrong creature could easily end up dead. Once a bear grows old enough to look after itself, the next form of aggression it encounters is when a mother chases it out of the territory or sisters begin the territorial chases that will affect where they will spend the rest of their lives. Later on in life, females will continue to use aggression, as we have seen, to reinforce their rules and their boundaries.

On at least ten occasions I found Yoda or Squirty treed with their cubs and witnessed the aftereffects of territorial aggression. I could usually tell when Yoda was treed, because her telemetry signal came in strongly from everywhere, so it was evident when she hadn't moved for a while. As I approached, I would call out to her and she would answer. I once found her up a spruce tree near the Black Brook blueberry patch. She had one cub above her and one missing. Two courses of sunken footsteps radiated out from the base of the tree–the results of a stiff-legged warning delivered by the adversarial bear whom I had probably scared off as I approached. Yoda came down from the tree when I arrived and immediately marked over the trail of sunken footprints with her own stiff-legged walk, releasing tiny drops of urine as her signature. Her message, in all likelihood, was that she planned to continue to use the food source. Then, in a very desperate search, she called out with a gulp-grunt to locate her missing cub. Finally finding her, Yoda sat at the base of another spruce tree fifty yards away with her nose pointing up toward the prodigal cub. This cub, still scared, chomped and huffed. It took her a while to come down.

I was also with Yoda on several occasions when the shoe was on the other foot: She was tracking and hunting down another bear. As I followed her along the Mascoma River in Dorchester, she located another bear's fresh scent on a red pine that she used as a marking tree. A raided hornet's nest lay on the ground, the displaced hornets were still flying about, and the trampled vegetation wasn't yet wilted–all clues that the interloper had just been there. It was a clear violation of Yoda's rules. She climbed to the top of the red pine and vigorously bit away, putting an exclamation point on her mark. Back on the ground with her two cubs trailing behind her, she began tracking her adversary step by step, marking over its scent with tiny drops of urine. I followed her until nightfall, and suspected by morning she would have the other bear up a tree.

What we see most often in bears is moralistic aggression, something evolutionary biologist E. O. Wilson associates with altruistic traits, like sharing. In his book *Sociobiology*, Wilson writes, "The evolution of advanced forms of reciprocal altruism carries with it a high probability of the simultaneous emergence of a system of moral sanctions to enforce reciprocation."

Long before I started researching the literature for an understanding of this behavior, I understood that the bears had rules that I was conforming to. If I stepped out of the wordless contract I had with them, I would be punished, and if I behaved I would be rewarded. Of course, it was a two-way street. Both the bears and I were engaging in robust manipulation–far beyond the begging kind I have experienced with dogs. The bears were exhibiting complex cognitive and social behavior, certainly not the kind of behavior expected in an animal that is considered by science to be solitary. Not only did the bears demonstrate a sense of quid pro quo, and the ability to set and enforce rules, but they also had clear means of settling disputes.

Just months after they are born, black bear cubs begin practicing the fine art of settling disputes. Their very first clashes occur when jousting for milk from their mother's bosom. Born with distinct personalities and gender-related differences, their contests of wills can lead to intense fights. With loud roars and focused, bulging eyes, Bert, Squirty's first son, deliberately struck and bit his sister Snowy's head, neck, and shoulders for dominating the nipples and the milk. With this action erupting on her belly, Squirty made no effort to interfere: Her cubs were learning the basics of fairness and developing the social skills that would carry them through their lives.

The first time I saw this behavior in the wild was when I crawled into the entrance of Squirty's den to film Bert and Snowy in late March 2000. Squirty had fashioned a toy from a sugar maple sapling in front of the den. She had peeled the bark with her teeth and presented the six-inch-long green stick to her cubs. The next day I brought her another toy for her to give to her cubs, a five-inch-by-one-inch piece of deer antler. Squirty promptly took the antler from me in her mouth and crawled back into the den to present it to Bert and Snowy. When I came the next day to film the cubs, I bribed Squirty with some kibble and proceeded to stick my head into the den entrance. I had a light that I propped up in the snow to illuminate the den. As I started to film the now four-pound cubs, Snowy took possession of both of the toys while Bert, who was not visible at first, emerged from a nook in the den. His eyes were large and his voice was fierce as he lunged at Snowy for hoarding both toys. Bert's squalls brought Squirty on the run, and I was quickly evicted from the scene. Even at this very early age, there was an expectation to share. When that expectation failed to come to fruition, it was met with an explosive emotional response.

By the fall of that year, Bert was noticeably smaller than Snowy; her ability to dominate her mother's milk had paid off. On a regular basis, I would locate Squirty with telemetry to make observations. As always, I brought a small reward of food, in this case a bag of kibble, in exchange for her time. I would spread it out on the ground and Squirty, Bert, and Snowy would share–that is, until there was just a small amount left. At that point, Bert would sit on the remainder, pin his ears flat to his head, and roar. Snowy would back off immediately, but sometimes Squirty would try to get some more. When this happened, Bert would swat her across the nose and she would submit. Again, with an explosive emotional response Bert was able to demonstrate that his need was greater than theirs; it was only fair. There was no retaliation on the part of Snowy or Squirty, they understood, and when Bert was done with his extra portion he rejoined the group. The dispute had been settled.

My observation logs have many such examples of bears using explosive emotional responses to mediate expectations of fairness. They were a frequent occurrence.

I wrote extensively about these in *Among the Bears*. At the time I wasn't confident of my analysis. Being self-taught, it was difficult not to be intimidated by the extensive literature in the behavioral sciences that clearly states that none of the behavior I had been witnessing actually existed in animals–or, if it did, it only existed in apes or humans. I'm over that now.

After observing and recording many disputes over many years, I have no doubt that they not only exist but also exist for a reason. Each bear is an individual with the basic desire to succeed in passing on its own DNA. Each bear has big teeth and big claws that are capable of doing considerable damage. Any social system where food sharing is an expectation is going to also have clear communication to avoid misunderstandings and communicate intent, and a mechanism to punish animals who fail to live up to expectations.

In short, I saw bears in the wild interact with the same unwritten compacts I had observed in my own relationship with Squirty. Our shared knowledge of each other's intentions, beliefs, and desires was based on each other's actions. We cooperated with each other based on our respective experience. We modified each other's actions with emotional communication and gestures of cooperation. Squirty manipulated me with what she had in her arsenal as a bear and I tried to manipulate her with what I had in my arsenal as a human. But always, in the end, it was Squirty who defined the limitations to my activities.

This was an individual relationship that would not have been possible without the mutual trust that developed during the fifteen months that I was her surrogate mother. She trusted and treated me like part of her family. But that alone was not a passport to her adult world. As an adult bear, Squirty limits her trust to me and only me despite the fact that my wife and sister helped raise her–clear evidence that our friendship had to be maintained with our cooperative interactions and mutual expectations. I had to conform to her world, not she to mine.

When I lecture about bears, an often-asked question is: You raised these bears as a surrogate mother, aren't they going to like *all* people? Some of the bears I've raised from infancy have had an initial curiosity about strange people they met in their transition to the wild, but they don't get the same signals from strangers as they did from me and interest quickly passes. They also learn very quickly that they cannot approach strange bears. It is no different from a human child who is quite trusting of strangers when young but, with experience, learns to be wary. The same is true with ravens and other wild animals I have had relationships with: Once they experienced the unpredictability of strangers, it became more difficult to reconnect with old friends.

Buried in that question, though, is a curiosity about whether the human connection causes Squirty to act differently than a wild bear otherwise would. My relationships with Squirty, Yoda, and the other bears we raised were novel. It is one thing to observe bear behavior through a pair of binoculars and quite another to experience their behavior firsthand in a give-and-take relationship. But even with bears raised so close to humans, it's safe to assume that the behavior they demonstrated in their interactions with Debbie, Phoebe, and me was their own.

Even my interactions with Squirty's offspring were different from my interactions with Squirty. There was a level of trust, but they did not have the intimate knowledge of me that Squirty did. They treated me with a great deal of caution and were clearly intimidated by my presence despite the remarkable level of trust that they had inferred from my relationship with their mother. In short, I was known to them, but I was not one of their cooperators. Still, I could observe their carefully managed relationships, systems of reciprocity, and competition over fairness and access to resources. I watched them as they combined aggression and reconciliation to build and maintain coalitions that would ensure their mutual survival. And in the end I've seen that they and other wild bears seemed to operate the same way Squirty did–both with other wild bears and with me.

There is nothing simple about this type of social behavior. Managing an equitable distribution of unequally distributed resources is complex business. It involves building coalitions with relatives to directly promote one's own genes, and developing cooperative relationships with both friends and strangers in the hope of a reciprocal acquisition and access to resources at a later date or in a time of need. As it happens, due to both the uneven distribution of high-quality food in patches and the seasonal variability given environmental factors such as droughts, early and late frosts, succession, and general feast-or-famine food availability, the cost of this type of sharing is low and the chance to reciprocate is often.

This level of sophistication in the bear world comes as a surprise to many, but not as much of a surprise as findings about how bears think.

Chapter Six

Cognition: What Bears Know

If you have read this far in the book, you will know that the bears I have observed carrying out their daily lives have proven to me their ability to share and cooperate, judge and punish, forgive and reconcile. They have shattered myths, showing that they are social, not solitary; communicate with intention, not just emotion; operate with a moral code; and even demonstrate altruism, compassion, and empathy. How else would you explain events like Squirty adopting the weakened cub Josie from her then-overtaxed daughter, SQ2?

In other words, they are intelligent creatures with cognitive skills well beyond those science currently recognizes in bears or other nonhuman animals. That's not to say that *all scientists* hesitate to acknowledge these skills. There has been a surge of interest in demonstrating animal intelligence over the past two decades. Scientists have devised experiments to document the advanced language capabilities of dolphins, the problem-solving intelligence of crows, the toolmaking and accounting skills of chimpanzees, and more. Birds, apes, rats, cats, elephants, dogs, and more have been confronted with puzzles to solve, dots to count, shapes and colors to recognize, and effects to associate with causes.

The curious have wanted to know: Which if any animals possess insight and intelligence? And how similar is it to our own? Do some animals have the ability to interpret what other animals are thinking—something that in scientific lingo is referred to as having *theory of mind*?

Along the way, the public has grown fascinated by some of the animals at the center of this research. Alex, scientist Irene Pepperberg's famous African grey parrot, showed us that he could associate simple words with meanings, distinguish shapes, colors, and numbers, and exhibit other cognitive abilities on par with what we might see in a human toddler. We've even learned that animals can experience heartbreak, as when Jane Goodall recorded the young chimp Flint withdrawing in grief after his mother died, or when Joyce Poole documented an elephant mother mourning the death of her newborn.

Still, findings that make the case for advanced intelligence—like those that make a case for morality, consciousness, and empathy—are hotly debated, and science as a whole has not accepted the fact that actions perceived as, say, intentionally strategic or empathetic are indeed that.

As animal behaviorist Marc Bekoff and bioethicist Jessica Pierce point out in *Wild Justice*, even one of the most public displays of animal empathy in recent memory didn't convince everyone. In 1996, a three-year-old boy visiting Illinois's Brookfield Zoo fell over the enclosure wall of the gorilla display, landing twenty feet below, much to the horror of his mother and onlookers. Binti Jua, a lowland gorilla, came to his rescue, making it clear to the other gorillas that they should back off, gently lifting the unconscious boy, and carrying him to safety—in this case the enclosure door—where she carefully handed him to zookeepers.

The gorilla's goodwill made headlines across the globe. But as Bekoff and Pierce write, "there was considerable skepticism among scientists that an animal, even an intelligent animal like a gorilla, could have the cognitive and emotional resources to respond to a novel situation with what appeared to be intelligence and compassion. These skeptics argued that the most likely explanation for Binti Jua's 'heroism' was her particular experience as a captive animal." Zoo staff had trained her in the art of mothering with the help of a stuffed toy, and that training involved sometimes handing the toy back to zoo staff.

Debates like this don't surprise me, but they have also come to concern me less over time. With nearly twenty years of qualitative data from the field, I don't doubt that animals can think strategically, act empathically, use language (communicate) intentionally, or even make calculations. And while I devise some experiments in the field to confirm my theories of how bears think and act, I often find that what Squirty and other bears show me is far more advanced than what has been recorded in controlled experiments in zoos or labs.

Take counting. Numerous studies have been conducted on how animals count. Primates, for instance, have been shown to be able to count to three and then estimate approximate sets, if not specific counts of objects, thereafter. When researchers realized that most if not all of these counting experiments had been done using social animals, they set out to devise an experiment with nonsocial animals to see if the same held true. They picked black bears (another sign that the scientific community has not yet acknowledged the social life of bears) and demonstrated the same ability using zoo-based bears and touch screens.

Such reinforcements of bear intelligence are heartening, but I also know from interacting with Squirty that she has no trouble counting—and that she can count much higher than three. I used to offer Squirty a sleeve of twelve Oreos, one by one, as a treat. When finished, she would turn and walk away. I decided to withhold one and sometimes two. Each time she seemed to know and made it clear that she wanted the stragglers. If I didn't produce the missing cookies, she would look for them herself. I repeated this test often enough, with the same results, to have no doubt she could count to twelve.

Like anyone studying wild animals, I am deeply curious about what they know, and frequently surprised. But I didn't come to my long-term bear study with preconceived notions about how they lived or interacted. I didn't expect them to show signs of behaviors that we regard as uniquely human, or that had parallels in the human world. It was the bears themselves who have taught me otherwise. And that began when raising the cubs.

My learning curve was not unlike what human parents go through with their first child. Squirty, her siblings, and other cubs I've raised have all been a handful. They were smart and adventuresome, and each had a mind of its own. They were better at managing me than I was at managing them. In fact, the cubs themselves taught me all I know about raising bear cubs, and they did this with the skillful application of pain and pleasure–delivered via a dynamic range of emotional vocalizations, message bites, and creative gestures.

A human baby manages his or her parents with a similar application of pain and pleasure. A baby's cry is so annoying that we search desperately for the means to comfort him or her. The child cannot tell us what his or her needs are; only that something is needed. The punishing behavior continues until we figure out what the problem is and we solve it. Then we are rewarded with peaceful bliss, and it all seems worthwhile.

The cubs' needs were always greater than mine, and it was in their interest to take as much advantage of me as possible. It seemed the more I responded to their emotional demands, the greater their expectations were. This was especially evident with food. The more I gave them, the more they wanted. I understood that raising cubs who were too big was not necessarily good for the cubs. They would have to make major adjustments to compete for food in the wild, and I didn't want to accustom them to abundance. But avoiding this was not always possible.

While Debbie and I don't have children of our own, I have seen this same behavior in human children: The more they get, the more they want. It is a very difficult thing to deny the wants of a child. The little monsters are cute and smart, have minds of their own, and above all are persistent. They can wear a parent to a frazzle. Both cubs and kids need discipline, but how? I didn't know until Squirty had her own cubs, and modeled perfect discipline. I raised her, and I couldn't achieve anywhere near her level of control.

She achieved this also by combining pain and pleasure, but she did it with great skill. When she had to, she'd tree a cub with an aggressive move to protect it from other bears or to reinforce her position in the family hierarchy; but she was never more aggressive than she had to be to suit the occasion, and most interactions were loving. There are few scenes more heartwarming than watching Squirty play with her cubs. The tiny cubs would jump on her back and slide down over her head. She'd cradle them in her big paws and hold down the wiggly tykes as she groomed them with her incisors. These were tender scenes, but how did she manage them so well?

Some, but certainly not all, human parents understand how to effectively manage their child's behavior. When I was growing up, my mother did the day-to-day care. She was an exceptional mother: She listened to us, loved us, and managed our sibling fairness issues. But every once in a while we pushed her buttons with our whining, anger, crying, and persistence. That was when she brought in the big gun. My dad was like Squirty. He had an explosive and unpredictable roar that all of us feared and none could get a handle on. This anger appeared so infrequently, so unpredictably, and with such emotion that the mere threat of "I'm going to get your father" was effective in getting us to behave. Despite the fact that we could converse with language, our actions were managed with the emotions of love, trust, and anger.

Squirty, of course, went on to build a dynasty with her offspring, and she rules that with the same skillful use of pain and pleasure. When any of her offspring return to the clearing, where I can witness their interactions, and they get close enough to her, she chases them. It may be a short chase or once around the clearing before she gives up in a blustery barrage of snorts, huffs, and false charges. Her offspring are used to this and just circle back to continue feeding; she lets them. They understand that Squirty is simply maintaining her position at the top of her family hierarchy. But I have also seen Squirty engage in sharing that requires some level of decision making. In the years after she had been bred but was without cubs, she would relinquish the highest-quality portions of her home range to her daughters who had cubs. In other words, she was sacrificing her best food sources for her daughters and grandchildren.

I don't know what to call this other than altruism–a trait that many comparative psychologists argue nonhuman animals lack. How might such a cognitive skill have developed?

The notion that wild animals lack the capacity for altruism is rooted in part in the Darwinian view that individuals place more emphasis on competing to survive than on cooperating to prosper. But we've already seen that bears tweak this rule. It's not that they don't compete for resources. They do. But they also cooperate to manage their lives, especially with kin. In fact there are no better-known cooperators than family members, as they have much longer and more frequent interactions to establish rules of cooperative behavior.

Those rules, of course, involve reciprocity–something that emerges when individuals or groups of individuals are unable to control enough resources to survive and reproduce and are forced into cooperating with those who have control of surplus resources. The distribution and timing of these resources must change often enough for both parties to benefit from the relationships. Friendships and coalitions will develop under these circumstances that allow all individuals to access surplus resources wherever they occur on the landscape. When these conditions exist, cognitive mechanisms should evolve to manage a new set of social problems.

One of these mechanisms is what evolutionary biologist Robert Trivers has described as reciprocal altruism. But reciprocal altruism has only been described in humans—suggesting that cooperation was a conscious choice of hominins who had already evolved large brains and advanced cognitive behavior. Marc Hauser, in his book *Wild Minds*, explains the concept this way, using an analogy from the TV classic *The Flintstones*:

> Fred and Barney are out on the savanna looking for dinner when they run into a sabertooth tiger. Both Fred and Barney know they can kill the tiger if they fight together; neither can survive on his own. They must cooperate. There is, however, a cost associated with fighting. Although the sabertooth tiger will either run away or die, Fred and Barney could get hurt in the process. Consequently, both men will be tempted to escape before the battle ensues. This tension between cooperation and defection has been called the prisoner's dilemma . . .

What I've seen in bear society, though, suggests that the evolution of cooperation and reciprocal altruism occurred at a much earlier and simpler level—and that it is not restricted to humans. If I am right about this type of social behavior existing in bears, then there should be parallel cognitive behaviors that exist in bears and humans, even perhaps to a different degree than they exist in other nonhuman animals. Let's take a look at some of the evidence.

Some scientists argue that the reason reciprocal exchanges between unrelated individuals don't exist in nonhuman animals is that they lack, as zoologist Tim Clutton-Brock wrote in 2009 in *Nature*, "language and associated psychological capacities to establish the intentions and expectations of both parties regarding the nature and timing of exchanges."

Once again, Squirty's adoption of Josie challenges that notion. While I don't know exactly how it was achieved, I do know that a specific agreement took place between Squirty and Josie's mother, SQ2, when

SQ2 visited Squirty's home range the night before the exchange. My telemetry signals, as you might recall, indicate that that was the first time the bears came into contact. That agreement had a long-term result: From that time on, Josie was Squirty's adopted daughter. Over the ensuing year, the families met many times at the clearing. If by accident Josie got too close to her natural mother, she was false-charged, reinforcing the terms of the agreement.

For this agreement to take place, SQ2 needed to convey to Squirty that there was a problem. She had to know that Squirty was a possible solution to it, and that Squirty had the ability to empathize with the problem and be receptive to the adoption. SQ2 would have also had to weigh the advantages of a life with and without her daughter over a year.

Squirty needed to be able to recognize and understand what SQ2 was trying to convey. She had to understand that SQ2 wanted her to adopt Josie; and she had to agree to a long-term solution. Squirty also had to apply the request to her present situation and consider her ability to raise the extra cub one year into the future. She had to determine that while she already had a single cub, she had enough surplus resources and was in good enough condition to support a second cub. Squirty had to believe that SQ2 wished her to adopt Josie. They both had to understand that the adoption was in the best interest of Josie's survival.

For any of this to occur, three elements of cognitive skills typically reserved for human behavior had to be in play: reciprocal exchange, theory of mind, and recursion.

Recursion relates to our ability to remember a past event, apply it to the present, and project it into the future. It's widely assumed that animals' minds lack those abilities–in other words, that they function only in the present. For reciprocity to exist, one must be willing to exchange a favor at the present for a favor in the future; to do that requires a recursive mind. And if bears truly have such a mind, as the adoption story would suggest, then examples of it should be common.

Squirty's behavior suggests that it likely is. Just a year before she adopted Josie, she pulled another stunt that reflected her cognitive abilities. That's when she gave birth to the single male Cubby, and by all accounts was a good mother until the middle of the breeding season when she deliberately interrupted nursing long enough to induce estrus. I say that she did this deliberately because she had sufficient weight to nurse her cub throughout the year, suggesting she had other reasons to abandon Cubby. Let's revisit that story and see just how it illustrates recursive behavior.

Female black bears are known to abandon cubs during food-stressed years, but this had nothing to do with food stress. It appeared that Squirty wanted to mate again in the hope of producing another litter with female cubs (helpful in expanding her greater home range), or at least more cubs. She seemed to know that once she successfully interrupted nursing she would induce estrus, which would automatically attract male bears, as her condition was advertised on the airwaves of olfaction.

Squirty's plan worked. She came into estrus and mated. In the process she demonstrated consciousness, self-awareness, long-term memory of events in past years, planning, and recursion.

How so? She was conscious of the conditions that brought on estrus in the past–that is, she understood that when a male arrived and she abandoned her yearling cubs her nursing would be interrupted and she would go into estrus, allowing her to mate. She was able to apply that understanding to a current and novel situation of having only one male cub. And she appeared to apply her past experience to the present, while hoping for better results in the future.

Squirty is not the only female bear in good condition who has pulled this stunt. While lecturing around New Hampshire I have been made aware of two other situations where females who were not food-stressed abandoned single cubs in order to mate in consecutive years.

One of the best examples I have of recursion, though, was captured on film and played worldwide in the National Geographic documentary on my work, *Bear Man*. I was in a dicey situation, and the documentary emphasized the drama of my plight and missed completely the importance of Squirty and another bear's behavioral sequence. Of equal interest to me has been the fact that despite a viewing audience in the millions, nobody recognized the behavior for what it was. I suspect the discussions will begin only after I apply words to describe the significance of what has been captured on film.

Part of the film documents a several-day courtship between Squirty and a mate. As I was filming, Squirty would at times follow me back to my truck, leaving her mate behind. On the fifth day of the courtship, the male bear came uncomfortably close to me. I decided to leave, intuitively thinking that Squirty and her mate would like to be left alone. But as the film shows, I was surprised when the male didn't allow me to leave. If he were only acting in the present, he would have been happy to see me go. I tried again to leave, only to be stopped a second time. The male bear then made it very clear to me that he and Squirty would leave and I was not to follow. This suggests that the male bear remembered that Squirty had previously followed me when I left, abandoning him; that he had applied that memory to the present when I tried to leave; and that he projected her behavior into the future, predicting that she would again follow me if I left. The sequence is an excellent example of a recursive mind. It is also evidence of theory of mind. The male was inferring her thoughts, calculating that she would want to follow me if I walked away.

In this same sequence, Squirty defended me each of the three times the big male made advances toward me. This suggests that she understood his intentions and acted to prevent him from harming me. This, too, supports my argument that bears are intentional communicators and have the ability to read the beliefs and desire of others. Her actions to defend me from harm bring up another significant and relevant aspect of her behavior.

In protecting me from the big male's aggressive advances, she showed evidence of altruism. Of course she probably made the mental calculation that she could protect me from the much larger male with little risk to herself because he desired to mate with her. But it is not the only time she has risked her safety for my own. Her behavior also suggested gratitude and reciprocation. This indicates yet another recursive sequence: Squirty remembered the value of our relationship in terms of friendship (a contract), food, and trust; she recognized the male bear's aggressive intentions toward me (the present); and projected a future without our relationship and acted.

This example stands out for many reasons. I was able to film it for others to scrutinize. I was part of it, which gave me a rare insight. I raised Squirty and enjoyed intimate knowledge of her behavior. And the big male was truly wild, a bear I had no previous experience with. The result was a rare window into the complex world of the black bear.

The scientific community, particularly the comparative psychologists, will take exception to the fact that I haven't performed a controlled and repeatable experiment where I design a circumstance for captive animals to perform the same behavior and confirm the result. There is a great chasm between what the general public experiences in the cognitive world of animals and what the scientific world has been able to prove. One has to wonder if science sets standards and complexity to seek the truth or to exclude competition in the great race for funding, peer-reviewed publication, academic posts, and the like.

It appears that bears plan ahead when it comes to choosing a den site. I was with Yoda in July 2003 when she spotted a blown-down spruce that had fallen over an old rotted stump. I filmed her as she broke her way through a tangle of dead branches and started digging. She started in one spot but was unsuccessful; then she tried digging in the rotted stump and excavated a den large enough for herself and her yearling cubs. She dug vigorously with her forepaws, pushing dirt and rotted wood between her hind legs and occasionally pulling the pile back to make more room. As she dug, her cubs came in to investigate and to do a little digging themselves. The whole episode lasted about ten minutes, until Yoda and the cubs moved on their way.

Yoda did not end up using that den in the winter to come, but her behavior provides a cognitively rich example that suggests that she was planning ahead. It was only July, but she was already taking time to explore prospective den sites for use many months from then. It also suggested that she had preconceived criteria for a prospective den site.

Using my homemade GPS collar, I have also documented Squirty taking two long, three- and four-mile journeys in successive years to pre-determined den sites. In early October 2002, just when the leaves had dropped, removing the security of their cover, she made a direct nonstop trip to a den site halfway up Smarts Mountain. The collar produced a location per minute, or about fourteen hundred points a day, so I could tell her route was uninterrupted. There was no searching around for a secure location. It was likely that she had done her homework on a previous trip to Smarts.

The following year in late October, after the first snowfall of the year, I followed Squirty and her cub's tracks from Roy Day's tractor shed on the west side of Lambert, over Lambert, and up into the bowl halfway up Smarts Mountain. It took me six hours to hike up there and back. Her route was direct, and her travel nonstop until she reached her destination. She was not randomly looking for a place to den. She was taking a long trek to a pre-determined destination. Again and again, her actions suggested long-term planning based on prior experience.

Ethologist and ornithologist Niko Tinbergen once suggested that the reliance of the scientific community on the controlled experiment may actually "prevent" the understanding of a problem. In reality, the commonsense observations ordinary people have of their pets may in some cases be well ahead of the scientific world. I have observed advanced behaviors in both of the Labrador retrievers Debbie and I have had. Stuffy came to us after his owner, Bob Gillie, a friend of ours, passed away suddenly. He was eight at the time, and the experience was traumatic for him. After he recovered from the trauma, he began to act out a particular routine. He would solicit first, then hug with his forepaws, then kiss both Debbie and me–and he did this every day for the rest of his life. If we didn't respond or ignored his solicitation, he would get noticeably upset. We read this as gratitude for taking him into our lives.

Buddy, a chocolate Lab whom we adopted after Stuffy died, had some tricks of his own. I didn't get Buddy neutered until he was two, after experiencing the roaming problems associated with male dogs. Buddy had discovered a female Lab at Donnie Cutting's house, about half a mile away. Roy Day and I were using a bulldozer and scoot, a large sled designed to haul logs, to pull wood off the Bear Hill Conservancy property that I manage across the road from my house. It was a cold winter day, and getting the bulldozer battery to turn over took some time. Roy and I had set up a heater and hooked jumper cables up to my truck, and we were waiting for the dozer to warm up. Buddy had previously figured out that the snowmobile trail we were on led directly to Donnie's house, so we kept a close eye on him. He was being very good, checking out coyote scent on the trail and staying relatively close. As we talked, he moved farther and farther away while still acting like he was only interested in the local smells. I called him back once, but eventually he again slowly moved off down the trail while periodically looking back at us. We decided it was time to try the dozer–and when I looked up, Buddy was gone. He had taken off at a run as soon as he noticed we weren't paying attention to him. I retrieved him at Donnie's kennel. Buddy's behavior suggested both planning and deception.

How does all this relate to what we see play out in the lives of other animals? I would learn a bit more about that from Richard Wrangham, a primatologist from Harvard University who had begun his career studying chimpanzees with Jane Goodall at her study site in Gombe and currently had his own study site in Uganda.

I met Richard for the first time in October 1996 at the request of Richard Estes, who had done some of the early work on the vomeronasal organ. He joined me for a day in the field with Squirty, Curls, and The

Boy. From the moment we stepped into the woods I felt a kindred spirit in Richard. Here was a person who knew and respected what I was doing. We spent eight hours together with the cubs, discussing different types of behavior and experiences, and while we talked the cubs performed. At one point, we came onto a large crooked black cherry tree that wild bears had fed beneath. The cubs responded to the wild bears' scent with interest, then fed on the cherries that littered the ground. It didn't take long for them to look up and see there were far more cherries still up in the tree. The tree had shaggy loose bark and sloped in a favorable fashion until about halfway up, when it twisted around. First Curls, then each of the other cubs took turns climbing halfway up, assessing the situation, determining it to be too dangerous, and then backing down. Richard said that he had never seen that behavior while observing wild chimpanzees. They always just went right up the trees they climbed. I thought about his remarks and realized that bears might be more insightful because they have to constantly solve problems they have never encountered before. Chimpanzees live in territories, as family groups, and have the luxury of following known routines.

I can relate to this distinction personally due to my dyslexia. I have been solving problems with only my wits since I was a child. Always behind in school, and with reading not my primary source of information, I have relied on observations of reality, insight, reason, and logic for my success. I could understand how chimps would be more accustomed to relying on information passed on through others while bears would need to act upon firsthand observation much of the time, a process that involves trial and error. In fact, failing to understand this need for firsthand information, and the role of trial and error in learning, is one of the reasons that schools fail to adequately teach slower readers and those of us with dyslexia.

The dyslexic brain relies less on recall and more on observing the situation at hand and processing it into a sensible whole. We think in images and use them to assemble, in our minds, the ins and outs of how a system at hand works or a problem at hand can be solved. We could be encountering a sentence on a page, the workings of a turbine, or–as in my case–the behavior of another species. We use the images at hand to understand and experiment with our world, and we create new solutions because we don't always have access to the solutions that others learn conventionally. To us, error is not failure; it is a crucial element to making progress. This mental process has its advantages. It likely helped Albert Einstein and other dyslexic scientists, artists, and inventors think outside the box. If education were to stop trying to get dyslexics back inside the box, who knows what kind of creativity could be unleashed.

It was approaching dark when we reentered the cubs' normal home range–an area near their cage where they usually hung out and would not leave unless it was with me. They came over to me and started suckling on my fingers. I told Richard that that was their signal to let me know they intended to stay put there for the night. He was impressed, and felt they had planned to stay out for the night rather than go back to their cage. As we walked down the hill toward the sugarhouse, Wrangham gave me a very nice compliment: "Whatever you tell me I will believe."

He invited me to visit his chimpanzee research station in Kibale, Uganda, and to present my findings to his colleagues at Harvard. My financial position prohibited the trip to Kibale, but I did present my findings to his colleagues and graduate students. This was very early in my work, and I was struggling with the scientific terminology to describe my observations. I learned a few more words that I had to look up and think about, but the biggest thing I learned was the advantage of having colleagues with whom to discuss findings and share common understandings of the literature.

Like chimpanzees, they lived and worked in a group environment where much of learning comes from listening and following other members of the group. On the contrary, I have been forced due to my circumstances to build a step-by-step understanding in my chosen environments of engineering and biology based on my real-life observations. In the process, I have developed a method that allows me to observe, analyze, make predictions, and test my results with an acceptable level of accuracy at a minimal cost. My world was more like that of a bear–which needs heightened observation of the environment, insight, and reasoning to individually take on the challenges of everyday life. I began to understand why so much about bear behavior made sense to me.

I also began to understand how one could accept one mode of intelligence as the norm, and thus fail to recognize others. Such is the case, often, with animal intelligence. If we assume the great apes to be the most intelligent nonhuman animal, we will try to find their same behaviors in other animals. When we do, it will be considered a benchmark of intelligence. When we don't, we will tend to assume the animal's cognitive abilities are not just different, but less. We need to look harder, and Yoda's nest building one day showed me why.

Much attention has been paid to the tree-nest building of chimpanzees, gorillas, and orangutans; it is regarded as the ancestral beginning of architecture. It was originally thought to be an instinctive process, but more recently believed to be a learned behavior–a way to construct a safe sleeping zone away from predators. The practice is so common that scientists believe an ape may build up to fifteen thousand nests over a lifetime.

Similar tree-nest building has been reported in the Andean bear, the sun bear, and the Asiatic black bear, but there are few if any reports of the American black bear building a nest for sleeping. They do create inadvertent structures in trees as they break branches to feed on beechnuts or cherries. These may look like tree nests, but are not. They also frequently build elaborate ground nests or day beds comprising raked-up leaves. Squirty, for instance, has fashioned nests by harvesting and laying spruce boughs on snow-covered or icy ground. I suspect that the basics of this nest building are instinctive. Even the cubs I raised who had no natural experience with their mother manipulated the hay I gave them for their winter dens. The finer details seem to come with training, though. I didn't see any evidence of the cubs building elaborate day beds until after they had spent time with wild bears.

All of which is why I found Yoda's behavior one summer day more than intriguing. We were walking on a hot, humid, and buggy afternoon. The deerflies were swarming on Yoda and working the blood spots they had made in her ears. As we walked, she spotted a large leaning red maple tree and proceeded to climb. When she reached the forking branches, she started biting limbs and breaking branches inward to construct a comfortable platform. I filmed as she fell asleep well above the fly zone of the pestering deerfly. This was the one and only time I had ever observed and was lucky enough to film a black bear building a tree nest for sleeping.

If we think about this, it was an extraordinary event. Yoda was able to envision a solution to her problem. She had the insight, planning, and imagination to climb high into the canopy, construct a comfortable nest for sleeping, and get away from the bugs.

I have also found that bears can use imagination. We all know that human children routinely engage in pretend or symbolic play. Symbolic play has also been observed in the great apes by Cathy Hayes, Jane Goodall, and Richard Wrangham, and described by Marc Bekoff in canines and in dolphins. We observed it in a bear named Teddy, who came to us in May 1996. He was small, weighing about five pounds, and was the only cub we raised that year. Single cubs are more difficult to raise as they like company and often get attached to caregivers. To keep him company, we gave him a teddy bear. He took the teddy bear with him everywhere and slept with it. Through the spring and summer we walked Teddy in the woods, and in July we let him out to live in the forest and supplement his food supplies. He could come and go as he pleased, but he preferred staying out. When he left the cage, he took his teddy bear with him to all of his bed sites; it stayed with him until it fell to pieces. The teddy bear appeared to suffice as a symbolic sibling, but that was not the end of it. As Teddy neared the time he would be on his own, he found a favorite log that he used as his stand-in love interest.

Cubs with siblings have a fairly ritualized form of play that starts with them facing off on the ground with open mouths and quickly becomes a bipedal wrestle reminiscent of a waltz or perhaps a sumo wrestling session. Some biologists refer to this as mock fighting because of the aggressive looks on the bears' faces, but it is done in complete silence and I have found it to be a mechanism of soliciting friendship and food sharing. Using a remote video camera baited with a small pile of corn, I have filmed this same behavior in adult wild bears. On one occasion, I recorded a big male come in and start to feed; then a second large male came into view and circled him. Honoring his position, the second male came back toward the first and opened his mouth. Then the first bear, too, opened his mouth to share breath in a ritualized way, which may allow bears to identify each other at close distances. Then they stood up and wrestled–often a way to bond and build trust–before lying down and sharing that small amount of corn.

In essence, this is the same thing Teddy was doing–but with an imaginary sibling that was, in reality, a small balsam fir just about his height.

When I go back and read my earliest field notes, I am reminded that sorting out how and when bears mark was a huge priority. All three sets of cubs were out by themselves much of the time, and it was essential to learn how to tell when a wild bear was around or discern whether or not a sign we came across on our walks was made by the cubs or another bear. The bears, of course, had no problem telling whose scent was whose–whether bear or human. When Houdini was on his own as a subadult, I had received reports of him in Enfield, a town about twenty miles south of Lyme. Like other wild bears in the area, he was attracted to food in people's yards, but unlike them he wore a radio collar, so I received calls when he was spotted. After one such call I drove down to Enfield to track him. I located his signal in a wooded area near a friend's construction project and stopped to ask if anyone had seen him. They hadn't. Then, while we were talking, Houdini appeared from the forest and came right over to me. His ability to recognize my scent was good enough to be sure it was me—even twenty miles from where he had known me, and while I was with a group of strangers.

Could they recognize themselves with equal accuracy? I was comfortable that bears could recognize their own scent and therefore had self-recognition, but I was intrigued by the amount of fuss the scientific community was making about a mirror test devised by psychologist Gordon Gallup to assess self-recognition. He was marking animals with odorless dye, then leaving them with a mirror and filming their behavior to see if they showed signs of understanding that the mark they were seeing on the animal in the mirror was the same as the mark on their own body, and ultimately recognizing that they were the animal in the mirror.

I liked the idea of experimenting with mirrors; they were in my budget, which was fairly close to zero. I wanted to see how bears would respond to mirrors knowing in advance they had the advantage of recognizing with their advanced olfaction who the bear in the mirror was. I also wanted to gain an insight into the minds of other researchers who were doing similar experiments with other animals. Let's take a look at how they performed.

I was unable to duplicate Gallup's mark test as I didn't have captive animals, I wasn't willing to sedate, mark, and relocate wild ones to a captive test site, and I lacked a controlled facility with video equipment that would allow complete documentation of results. I had to proceed with what I had: free-ranging and wild animals that could explore mirrors or not. The cubs' first exposure to mirrors was very similar to what Darwin described when he placed a mirror in the cage of an orangutan at the London Zoo: They were immediately attracted to it. They looked at their images, then looked behind the mirror. They looked at their images again then looked beside the mirror. They mimed in front of the mirror, making paw-to-paw contact with their image. They also checked their image with their tongues for scent and showed curiosity, but never any fear. Curls, after she had left my care, returned to the cage, removed the cover from the mirror, and swayed back and forth in front of it. Houdini, after eating beaked hazelnuts, picked up the husks and repeatedly dropped them while watching his image in the mirror. I placed a collar with scent from a wild bear on it in front of the mirror and Curls responded to the scent, sniffing and licking to identify it. When she looked up and saw her image, she spooked and treed, then returned to the mirror to look behind it and further investigate the scent on the collar. Her response reflected a momentary thought that the bear in the mirror might be a stranger.

Later on, I had Roy Day build me a box with a two-way mirror that I could film via a remote camera triggered by a passive motion detector. Roy had fun when his friends would come by and ask what he was building. "Take a look, there is a monkey in the box," he'd say. When he finished it, I placed it at the clearing and baited it with a small amount of corn. The wild bears who came across it were initially surprised by their images, then slowly got comfortable enough to eat in front of them. SNLO, as a yearling, initially made quick run-bys, slowly gaining the confidence to confront her image. Eventually, she would mime, swat, and make open-mouth displays at the bear in the mirror. Squirty would feed comfortably in front of it.

I added a second mirror to the clearing, this a four-by-eight-foot stationary one. I faced the two mirrors at each other and placed the food between them. In order to eat the food, bears would be faced with images within images of themselves on both sides. This did not elicit a response at all; they quietly ate their food.

I also made an attempt at marking Squirty for the producers of *A Man Among Bears*, another documentary, produced for the National Geographic Channel. Squirty did not pay any attention to the sticky mark, which I applied while petting her, but she did remove it shortly after feeding in front of the mirror. This suggested that she was aware of the mark, but she also may have felt me applying it.

The final stage of my experiment was to leave the mirrors unattended and without bait to attract the bears to their images alone. I wished I had the funding to reliably monitor the mirrors, but I didn't. I was pleasantly surprised at the response by the bears over the course of several summers and found that I could monitor the use through the number of muddy footprints of bears as they stood bipedal and mimed in front of the mirror. I was able to film and photograph both Squirty and SQ2 taking their cubs to the mirrors to experience their reflections as well as the yearlings returning to the mirrors after leaving their mothers. I have a series of photographs of Squirty lying down comfortably in front of the large mirror while Josie and her son mime and check the scent of their images with their tongues. It is clear to me that bears have both olfactory and visual or a multi-modal self-recognition.

These findings support and are predicted by my overall model of black bear social behavior. Behaviors like these do not exist in isolation; they are just one link in a chain of behavior that was enhanced when bears were forced by environmental conditions to cooperate and communicate with strangers. The sense of self is a requirement in any negotiation that requires a cost–benefit analysis of a social or economic contract. To my mind, the fact that these contacts take place is prima facie evidence of the behaviors themselves. It is hard to make a deal with someone and not be represented in the interaction.

Another cognitive realm I set out to explore was imitation. Would bears learn from other bears by mimicking their behavior? Such social learning is the mechanism that allows for a rapid spread of information or culture through a population. It is how trends ignite in the human realm, and in the bear world it is how animals acquire new skills or information. Time after time, I saw how bears followed other bears to find food, and subadult males followed and learned from their adult male mentors. That mentoring process suggests theory of mind in that subadults are recognizing the advantage of following and learning from the more experienced adult. At the very least, it introduces the concept of knowledge as a resource.

At the 2011 International Bear Association Conference in Ottawa, Karen Noyce, a senior bear biologist from the Minnesota Department of Natural Resources, presented a paper that she had extracted from many years of telemetry data, showing that male bears follow each other over long distances. The study involved more than two hundred radio-collared bears over ten years and showed that 40 percent of the males traveled alone, but in the same direction, up to seventy-six miles to concentrated food sources. In the fall, 20 percent of the males moved between six and sixty-five miles to den in the expansive peatland landscape well north of their summer home ranges. Karen went on to say that the bears showed a degree of coordination in their movements, and she suspected they were learning from one another, following keystone individuals with knowledge, and using trail networks to find key resources. It was nice to see that Karen's study reached the same possible conclusions that I had from my observations.

Social learning not only helps animals find food, but also helps them know what to eat–such as when the cubs leaned down to check my breath when I was eating clover (something they hadn't yet learned to do), then immediately found and foraged on some. Squirty, Curls, and The Boy, in fact, learned how to eat jack-in-the-pulpit this way, too. Despite mouthing jack-in-the-pulpit, as they had done to learn the majority of their herbaceous food, none of the cubs had eaten it. One day Curls followed the hot scent of a wild bear who was feeding near their cage. She checked out twenty-four sites over six hundred feet where the bear had fed on jack-in-the-pulpit. After that that she was digging and eating it, too, and then Squirty and The Boy learned from Curls.

We, of course, are social learners, too. Sometimes, we even learn from bears.

It is said that Native Americans used to follow bears to learn which plants could be used for food and medicine. When first pondering this, I was stumped by why the Native Americans would decide to take their cues from a bear. But the answer came to me at a lecture I gave for the Vermont Fish and Wildlife Department. After I finished talking, a Vermont conservation officer came up and told me that his father had eaten skunk cabbage when they lived in Massachusetts but years later, while living in Vermont, he had pulled off a quarter-sized piece of a leaf of false hellebore, something that grows along the streams and in wetlands and is locally called skunk cabbage. His father became paralyzed from the waist down for more than a week. It dawned on me that, with experiences like that, it wouldn't take indigenous people long to find a better way to sample plants to see if they were poisonous or not. If there were pharmaceutical qualities to the plants bears were eating, they might have been taking advantage of them as well.

I wish I had direct evidence of a bear treating itself with plants when it was sick, but it is very difficult to know when a bear is sick, and in my experience, they do not get sick very often. Yet I did observe bears treating themselves on a number of occasions. When Little Girl had an impacted canine tooth, she shredded a spruce stump to dislodge it; when a fighting scar on Yoda's nose became infected, causing dramatic swelling, Yoda dug a wet den in a sphagnum swamp and spent five days in it to fight the fever from the infection. Bears also treat themselves with blue-gray clay, organic soil, decayed leaves, and ungulate scats. I can only speculate why, but evidence from other species suggests they eat soil, leaves, and scats for their probiotic properties, and the blue-gray clay to remove toxins from their systems.

After the first cubs I raised emerged from their winter dens, they ate large quantities of white pine needles. Their scats were entirely full of needles. I don't know for sure but do speculate that they ingested these to rid their intestines of parasites. Bears get roundworms from their mothers, and while the worms aren't very pleasant to think about or look at, they don't do any harm to healthy bears. It may be that after hibernation, though, when the cubs' intestines were completely emptied, the worms irritated the cubs' systems.

In *Among the Bears* I spoke often about the cubs eating fresh deer scats or droppings on the walks we took, speculating that the benefit to the cubs was that the scats were a source of organisms to help them digest vegetative matter (which was 85 percent of their diet). So I did an experiment by collecting the fresh scats they were ingesting and the older scats they were refusing. I placed samples of them in Ball canning jars, two containing the fresh scat, two others the older scat. Then I added water and the fresh beech leaves the bears were eating at the time, wrapped the jars with heating pads, and heated them to about a hundred degrees Fahrenheit, the bears' body temperature. After ten hours I checked the jars and found that the leaves I had placed in the jars with the fresh scats were completely dissolved except for a skeleton. The leaves in the jar with the older scats, however–which the cubs were rejecting–were fresh and waxy just as they'd been when I put them in. That experiment told me that there were organisms in the fresh scats that could digest cellulose, the protective covering of the cell walls that protect nutrients from being digested, but it didn't tell me that the cubs needed them.

I didn't get evidence of that until we received an emaciated cub on April 1, 1996. She was sixteen months old and weighed just eleven pounds. We fed her lamb's milk replacer, baby cereal, and Kibbles 'n Bits dog food, all high-quality foods for young bears. But by May 1, she still failed to thrive and had lost weight. Early in May when the deer were eating spring grass, Phoebe went out and collected some fresh deer droppings from Walt Record's hay fields. We offered them to the cub, and she gobbled them up. Within days we could see an improvement, and we released her in August weighing over sixty-five pounds. I have shared this information with other bear rehabilitators and zoo professionals, and ungulate scats are regularly used as probiotics for bears. I would like to add that our Labrador retrievers Buddy and Sophy also eat soil from selected spots and savor deer droppings when they find them, so it is likely that ungulate scats could be used to treat other animals as well.

Bears and dogs are not the only ones to treat themselves with animal dung. During World War II, the German troops in North Africa were dying in alarming numbers because of dysentery. They wondered why the Arabs weren't affected by the same problem. An investigation found that they were, but when they developed the first signs, they would follow either a horse or a camel around until it left a feces, then quickly grab a small amount and swallow it. The dropping had to be still warm to have an effect. The Germans isolated the *Bacillus subtilis* bacteria from the camel and horse droppings and cured the dysentery epidemic among their troops. *Bacillus* capsules are still available today to treat intestinal problems.

Just how much of bears', and other animals', self-prescribed treatment is the result of social learning and how much is instinctual know-how remains unknown. But this much is clear: In our quest to find out what animals know, we often forget to ask what they can teach us.

Chapter Seven

What Bears Can Teach Us

about Our Own Past, Present, and Future

It turns out that bears may be able to teach us quite a lot. While Squirty was my widest window into the bears' world, it might be that the bears' world provides a large, unexpected window into our own—albeit our prehistoric one.

To explain why, I'm going to go out on my biggest limb yet, and raise some questions about what kind of early pre-human behavior might have led to our own behavior in modern times. The advent of language forty thousand years ago, writing five thousand years ago, and more recently the telephone, television, cell phone, and Internet have all led to increased social control and changed the social dynamics of our ever-increasing population. But long before all this, it's quite possible that pre-human and early human culture and communication looked a lot like a bear's.

Could their food-sharing behavior show us how cooperative behavior between unrelated animals might have begun–and how it spread rapidly through populations? Could a close look at the primitive social systems of black bears provide an alternative explanation for the development of human cooperative behavior, altruism, and morality?

I think there's a very good chance that it can. Bears make a good model for early human behavior and social development because there exists in bears the component parts that would lead to human prosocial development, meaning the development of positive traits like compassion, morality, and the ability to help one another out through sharing and other basic means.

Consider yourself forewarned that this could be a controversial notion. And know, too, that what I posit here is just a gateway–an invitation to a new line of research in the quest to solve puzzles that have long persisted about how we humans came to behave as we do. But come along with me, and let's speculate about the possibilities.

First let's revisit the reasons why bears share. Bear society, evolutionarily speaking, operates at a very basic level. Bears don't form social units the way chimps, for instance, do. Once they leave their mothers, bears are more or less on their own, lacking the protection and support of a tight territorial group. Nor, as we've seen with Squirty, are bears able to control all of the resources they need to survive and reproduce. They would like to, but they can't: Their food is unevenly distributed across the landscape, not easily stored, subject to huge surpluses and shortages, available in patches, and limited to natural production. In other words, like our earliest ancestors, bears need to find food daily; they can't reap it from stored caches, and they certainly can't grow it. Female home ranges are defended and evenly spaced on the landscape. And at any given time, one female may have a surplus of food and other resources while another female may have a shortage or, for that matter, none.

Such obstacles set the stage for the evolution of social exchange and reciprocal altruism among relatives and non-relatives alike. Given these conditions, it doesn't seem surprising that black bears have evolved a social behavior that allows access to food and other resources for all individuals, even strangers. They form alliances and coalitions with unrelated bears and they share food based on a system of reciprocal altruism. They establish rules that define the parameters of this sharing, and they enforce them through a system of judgment and punishment. As bears evolved and their populations became denser, the need to share would have grown even greater, with a corresponding rise in the need for fairness and justice.

Our early ancestors, like today's black bears, were large omnivores who filled a similar niche in the ecosystem: They ate primarily the highest-quality food available. They would have had to travel to get what they needed. As their populations grew, they would have had to develop ways to access the food supplies claimed by others. Like bears, it's likely that they had no choice other than to cooperate with one another to get what they needed, when they needed it. To do this, they would have needed to develop the basic moral behavior that I've repeatedly witnessed in Squirty, her family, and her neighbors. That behavior would have grown exponentially more complex as technological success continued to grow our food surplus, populations became denser, and the need to interact and cooperate with more people opened the door to more problems of perceived cheating–and thus to a greater need for judgment and punishment. In the range of human societies that exist today, from tribal to cosmopolitan, this need for increasingly complex social controls plays out in norms, rules, and laws.

Bears' social behavior demonstrates that an early form of social exchange, reciprocal altruism, or cooperation between unrelated individuals could have evolved in our ancestry as early as the split between our direct ancestors and chimpanzees six to seven million years ago. Had the social exchange begun at a time when our ancestors were no more advanced than the bears under similar environmental conditions, our ancestors would have had a huge evolutionary advantage over the animals who evolved into chimps–explaining, perhaps, why our branch on the family tree led to modern humans.

This view of the roots of cooperative behavior differs dramatically from modern thought. Scientists currently speculate that social exchange began much later, in the last million years during the time of *Homo erectus*, an already large-brained ancestor. But that doesn't explain what put us on course to grow such large brains or why the chimpanzees did not.

Our path is lined with extinct ancestors leading up to *Homo erectus*. Based on my own observations of bears, it seems far more likely to me that social exchange occurred at a much earlier and simpler level and that it fueled our cognitive leaps, giving birth to language and increasingly complex social codes as the need to communicate with strangers came into play and gradually increased.

Recent findings on our early roots may back that notion up. Scientists have long thought that chimps and humans had a common ancestor until about four million years ago, when they parted ways on the evolutionary tree. But new evidence suggests we might have had an even earlier common ancestor, *Ardipithecus ramidus*. The physical traits of that creature have led scientists to speculate that it behaved quite differently from chimpanzees. For one thing, it lacked the large canines that male chimps use to fend off other males. This suggests that *Ardipithecus* likely wasn't a warrior; and if it wasn't a warrior, it likely didn't live in tight social groups, like chimps, that heavily defended a territory. It's probable that humans didn't, either, until we began to settle, create wealth, and fight to defend it or win more.

Scientists have searched out many models to examine the earliest roots of our social nature. They have found some, but key mysteries remain. We don't quite have all the clues we need to assemble the jigsaw of how the stage was set to make us, behaviorally, into the creatures we are today.

Perhaps we've been looking at the wrong models.

For the most part, scientists have looked for clues in the behaviors of creatures most like ourselves, genetically: chimpanzees, gorillas, orangutans, and bonobos. As a result, many support a theory of group social behavior that suggests that early human social groups looked a lot like the social groups we see in today's chimpanzees. This doesn't explain, though, how humans began cooperating with strangers.

Chimpanzees live in social groups dominated by a male hierarchy and consisting of many subgroups. They maintain tight family bonds, and males staunchly defend territories. It seems a mistake, though, to think that a group society like this could easily evolve into the openly cooperative and dynamic society that exists in black bears and humans. It's plausible that group societies could form out of an open society like Squirty's if the habitat conditions changed and allowed groups to dominate all their resources. But it is much more difficult to imagine the reverse.

The emergence of chimps as a model likely occurred because their brain-to-body ratio is about the same as most of our extinct relatives'–and also because they use tools, which together with cooking is thought to have fostered the success of humans in evolution. But even tool use raises issues about the chimps as models for early human social development. Though chimps do show widespread tool use, not all chimpanzee groups use the same tools. This fact suggests that their ability to broadcast what they learned across their population had limits imposed by their territorial group structure.

When we look to black bears, however, we see that the advantage of cooperating and communicating with strangers would have led to widespread cultural exchange throughout the population, and hence the rapid spread of any cultural advances. The chimp's limited exchange due to group territoriality has, in my mind, been one of the reasons chimpanzees, despite their high levels of intelligence, are still chimpanzees.

Bears, of course, are still bears, too, but their basic social behaviors hold important clues about how the kinds of social systems that parallel our own could have developed. We have seen that female bears forge reciprocal relationships with both related and unrelated females. Male–female coalitions form to access food in new areas, and to allow subadult females access to food in adult female home ranges. Coalitions of unrelated males form to access surplus food in female home ranges. Males are able to travel far and wide, learning of new food sources or possibly cultural advances. When food is scarce in one bear's home range, the bear can travel or follow others to learn the location of surplus foods. Complex networks of friends, social relationships, and communication become the fabric of a greater society.

If we look at why and how subadult males form and use coalitions to access surplus foods in female home ranges, we find a particularly intriguing model of early group formation among unrelated individuals. The power of these coalitions is limited by the availability of naturally occurring resources and is driven by the need to access surplus foods to survive. That surplus food exists in the home ranges of females, though. Our GPS data indicate that adult females cover the core of their home ranges every two or three days–a sign they are reinforcing their dominance in their territory, not just searching for food. They need to successfully establish and maintain a home range of high enough quality that they can have the resources required to raise cubs when the cubs are too young to travel. It is not in the interest of the adult females to share with the subadult males. The males have less need than other females, as they are not burdened with cubs. And they have less ability to reciprocate. So subadult males must find power in numbers to enter the territory of the adult female and access surplus food supplies.

In my study area, in pockets of surplus food, it has not been unusual to film eight to ten males responding to the small amount of bait in front of a remote camera. On more than fifty occasions I have taken pictures of two males sharing that small amount of corn, an indication of friendship. The DNA results of these bears show no genetic relatedness.

We see in these males how reciprocal friendships form, then transform into coalitions of, or group formations among, unrelated individuals. All this is triggered by the uneven distribution of resources on the landscape.

So even at this basic level of cooperation, population-wide norms have evolved to enhance the ability of every individual to access food and other resources. Young animals routinely test the tolerance of more dominant bears to learn the rules. The subadult males can expect that unguarded food can be taken without reprisal, but as they grow older and strike out on their own for mates they learn that local and individual rules must be followed, because they are routinely enforced. To do all of this, they need to interact with strangers, which leads to a whole different set of communication and social skills than communicating with family members.

Why, you might wonder, would there be such a difference between what it takes to communicate with a stranger versus communicating with a family member? Jared Diamond, in his book *Guns, Germs, and Steel*, describes the importance of being able to communicate with strangers. In a discussion about early human group formations relating to bands and tribes, he described two New Guinea warriors meeting on a trail and sitting down to a twenty-minute discussion of what tribes they were from and who they were related to, just to justify not killing each other.

It is interesting that both bands and tribes are egalitarian in nature—that is, they have no formal means of government, and they rely on reciprocal exchanges, not unlike bears.

Studying the way bears behave may answer questions not just about our past and present, but also about our future.

Animal brains and human brains evolve to solve problems. We first perceive the problem, we think about it, then we come to a conclusion. Animals perceive real-world problems firsthand and are likely to come to the correct conclusions. If they don't come to the correct conclusion, there is a direct consequence of decreased fitness. For most animals living under natural conditions, the Darwinian principle of survival of the fittest applies directly to the individual. I say this because in a wild situation all of an animal's cognitive abilities are needed for survival; there is little margin for error. If the right choices are made, one survives and mates. If not, selection takes place.

As humans, some of our conclusions are right and, more often than not, some of our conclusions are wrong. I can say this because unlike wild animals, our technology has given us such an increase in our fitness or ability to survive that there is no longer direct evolutionary control of our individual decisions. As long as our technology produces more food and other resources than we need to survive, there is no direct selective force pressuring the accuracy of our choices. As a result, our cultural archive is riddled with inaccuracy. At every level of our society, we can and have made bad decisions without any immediate consequences. We have made bad decisions that appeared to be good at the time and have omitted good decisions based on the bad information we had at the time.

Natural selection on the individual level is replaced with a selection of ideas and ideology based on what was known at the time. Because of this, consequences have shifted from an individual level to a species level. Competition of ideas, ideologies, and social control of populations and resources have put humanity on a rocket course of change. We are in a race to maintain our technological advantage and to assume social control of our spiraling population. The complexity of social control continues to increase as our population grows. We are on a fast track to either great evolutionary success or colossal failure.

Our short written history reveals this pattern of change as rural cultures have been dominated by urban cultures in search of resources, spreading ideology and religion in the process. The unwritten rule *Might makes right* has repeated itself throughout our history. What has been the basis for such behavior? Have we been making informed decisions or have we been making decisions that seemed right at the time? Are we under divine guidance or have we misunderstood the role of religion? These are all huge questions that I feel can become much clearer to us if we understand a natural basis for our behavior.

Again, let's look at the consequences of our decision making. The conclusions we come to that are right produce a benefit and persist, while those that are wrong or outdated ultimately have fallen by the wayside, and sometimes not soon enough. The point is that we need problems to develop to succeed. When there are no longer problems to solve, our evolutionary development will slow to a crawl. This in all likelihood has already taken place as we have transferred many of the drivers of our own evolution to technological advances.

If the need to communicate to manage social complexity led to language and brain development, then we have recently been in the process of transferring these drivers of intelligence to the technology of telegraph, telephone, television, radio, and the Internet. The question is: Will the human brain atrophy as a result?

Stable species of plants and animals develop to the limits of their niche. A niche is simply the space in which an organism has been able to develop within the environment. Each organism does all it can to expand the limits of its own niche. It competes and benefits from other organisms, bringing stability to its position. It also adapts in any possible way to survive in concert with the other organisms that it interacts with. Over time, this interaction results in stability or equilibrium within the vessel of earth's environment.

Throughout the history of the earth there is evidence of many disruptions of this process that have led to minor and major extinctions, resulting in new opportunities toward stabilization. These renewals have resulted in our own existence. Whether you believe in divine guidance or a chance collision of social flexibility and technological advantage, an experiment is taking place on earth with gigantic ramifications to the human species. We have been given the opportunity to succeed–in other words, to inhabit this earth for perhaps millions of years to come–or to be just a blip in the history of the eternity of life. The living system in the terrarium we call earth has no conscience;

it just doesn't care whether we carry on or become extinct. The living system that exists on earth has only one goal, and that is to be stable and reach equilibrium. With time, life on earth is like energy–a force that can be neither created nor destroyed; it merely adapts and evolves to continue its existence.

Let us leave the big picture for a moment and try to understand our own situation. In the process of evolution, problems are solved either with physiological or mental adaptations. Examples of this principle can be seen throughout nature. It explains why some insects can make and use tools, why a beaver can modify its environment so skillfully, and why the closest relative to man is still a chimpanzee. For any evolution to take place, there must be new problems to solve.

What I am saying is that if life on earth ever reaches equilibrium, each organism will develop and stabilize as it reaches its own limitation. That limitation is the point where an organism no longer has the opportunity to change or adapt, where it is able to stabilize and hold on to its place within the system of life. Once stabilized, an organism can remain in a stable form for millions of years. Gray squirrels, for example, are thought to have been in their present form for as many as fifteen million years; horseshoe crabs, one hundred million years.

For humans, technological success or tool use has produced a surplus of food and other resources, but it has also led to increasing populations, and by doing so has created a host of additional problems to solve. Solving the technological problems has been the easy part; dealing with the social ramifications of that technological success has been a source of seemingly unlimited issues. For every giant step made for humankind, we find ourselves closer to the brink–due to conflicting cultures, misperceptions, and competing interests.

Over time our brains have responded to these new social problems with increased verbal communication and ultimately language. In "On the Limits of Natural Selection," published in the *North American Review* in 1870, Charles Darwin wrote that "the intellect must have been all-important to [man], even at a very remote period, as enabling him to invent and use language, to make weapons, tools, traps . . . whereby with the aid of his social habits, he long ago became the most dominant of all living creatures. A great stride in the development of the intellect will have followed, as soon as the half-art and half-instinct of language came into use; for the continued use of language will have reacted on the brain and produced an inherited effect; and this again will have reacted on the improvement of language." He went on to observe, "The higher intellectual powers of man, such as those of ratiocination, abstraction, self-consciousness . . . probably follow from the continued improvement and exercise of the other mental faculties."

So it is that language led to better technology, which led to increased human densities, then written language, more people, and better technology. Yet we still haven't been able to contain the manifestation of nature that existed and works so well in Squirty's world: perception, judgment, and punishment.

It is my position that all of human behavior will, in time, be traced back to a simple and natural beginning. In my view, the simplicity and directness of bears' behavior toward the few individuals they interact with, along with the austere nature of their lives, make them great models for that exploration. But the assumption that we began our social lives much as bears operate theirs now might also show us how surplus fitness and an increase in population density have affected human behavior.

For me, it is quite sobering to realize that altruism, cooperation, judgment, and punishment are inextricably related and that they existed long before, perhaps millions of years before, humans had the opportunity to define their own existence. Why? Because we are behaving just like a bear in a world that has become so complex that none of us truly understands the basis for our own actions. We are active players in the lethal game of evolution, yet our individual fitness has nothing to do with our ultimate survival. We use the same time-tested behaviors that the bear uses to survive, only theirs are tied directly to their individual fitness. If they make a poor decision, they experience an immediate and perceivable loss.

The value of knowing how a primitive social system works, one that has a structure similar to our own, gives us a basis to judge our own actions. It allows us to predict and understand emotional responses to actions and look for root causes rather than taking action based on our own emotional volatility. If we understand that we all have an instinctive emotional barometer, one that was designed to work with reliable information and result in direct consequences that affect our individual fitness, and that this emotional barometer is still operating and regulating social behavior in a world that is complex, poorly understood, and dominated by a vast array of power centers and conflicting interests, then we have an opportunity to look past actions driven by sweeping emotion and poor or incomplete perception, and look for and identify root causes of problems. Put another way, understanding how a system of cooperation and goodness worked at its conception gives us an opportunity to look beyond the emotional drivers of conflict and get down to the rational causes of problems.

Warfare is the evidence that human social behavior acts and reacts without having any true understanding of the behavior itself. If we recognize the social behavior of the black bear–or any other animal whose social behavior might give us insight as a primitive model of human behavior–we can see evolutionarily stable systems that have existed for millions of years. In only the last two million years can we see the destabilizing effects that success has brought upon our once stable social behavior. With our large brains and ability to communicate, we have been given the opportunity to control and manage our own destiny. Unfortunately, we have a primitive and instinctive social motor that continues to function as if nothing has changed since the dawn of man. Warfare is a manifestation of this primitive behavior in the modern world. It doesn't exist in the bear's world because judgment and punishment and their consequences exist on an individual level. In my opinion, we were at that same level until about two million years ago when our ancestors started solving problems with tools that enhanced their collective fitness.

Elements of warfare do exist in the group social behaviors of chimpanzees. They have been observed beating and even killing members of neighboring groups. But in a bear's world, group formation of unrelated individuals is limited by the availability of natural resources. The evolution of this behavior has gone through the step-by-step process over five hundred million years. The sudden appearance, in evolutionary terms, of a sustained surplus of a wide variety of resources in the last million years of human evolution–with an exponential growth in the last ten thousand years–allowed this instinct to take on a new form without any natural limitations. Groups of unrelated individuals have throughout our history formed to dominate every conceivable resource including water, land, food, education, religion–whatever might produce a base of power. Once an essential means of accessing and distributing natural resources, the ability of non-relatives to form groups has evolved into a potentially cancerous aspect of our social behavior.

Every one of us experiences the effects of this change every day of our lives. It explains why nothing makes sense, why there are big gaps in fairness, why politics are so inconsistent, why unions outgrow their usefulness, why our educational systems cater to certain students, why science has standardized methods that preclude others, why religions try to expand their influence, why there is racism and discrimination and much more. The important aspect of this is that we aren't this way intentionally; it is our nature. As groups form or boards meet, there doesn't even have to be an intentional plan to consolidate power and limit non-conforming members; it just happens.

We could draw upon a number of well-known injustices to underscore this reality. But let's take a look at one that, for me, hits close to home. When we look at the evolution of education, we can see that knowledge has been a resource of power. In this country's past, it was illegal to educate slaves and women; women are still not educated in some places in the world. It wasn't until the twentieth century that an adequate education was mandated for all citizens of this country, and even after that there were problems with segregation and gender equality. In 1983 the National Commission on Excellence in Education issued a report called *A Nation at Risk*, which revealed problems in what students were studying, how hard they were studying, and how much they were learning. In response, some have raised academic standards and launched new testing programs. "Others," wrote Chester E. Finn, Jr., in a 1989 issue of *Commentary*, "have enacted comprehensive education-reform legislation, which add to graduation requirements, decrease the average class size, require teachers to take literacy exams, require students to pass standardized tests, redesign teacher-licensing requirements, and much more." Nowhere did they consider the diversity of how different students learn and consider a complete "outside the box" revision of the American educational system.

When I was on the New Hampshire Special Education State Advisory Committee, I was asked why the system was losing children as early as age five. As someone who was flunked out of kindergarten at age five, I replied based on my own experience. I explained that while all brains are anatomically similar, they possess plasticity and adapt to the individual during development. Howard Gardner, a developmental psychologist at Harvard University, theorizes that humans possess multiple intelligences: linguistic, logic-mathematical, musical, spatial, bodily/kinesthetic, interpersonal, intrapersonal, and naturalistic. I prefer to consider this range genetic diversity, which exists in every living organism and is essential to the ability to adapt to changes in the environment. To put this simply, each one of us is born with a way of learning that works best for us as an individual. I have been successful only because I have been able to use my own methods of learning. Had people who learn like me been able to maintain the dominance they must have had before the written word, I would have gotten A's in school and passed tests with flying colors. I would have advanced degrees and a high-paying job; it just comes down to which group controlled the resource of knowledge.

Early in our education process, children are asked, more or less, to express their plans for the future. Our social training by and large produces a response like "I am going get a good education, have a good job, and raise a family." And they'll likely have, already, even as young as five, some kind of an idea about how they will embrace and be successful in life. In other words, they have an innate sense of how they excel.

The teacher responds, "Sorry, but you can't do it that way; you have to do it this way."

Right there the child is derailed, as I was. Not only does he or she have to learn the material that is being taught, but he or she also has to learn somebody else's method of learning. Believe me when I say this can be an insurmountable task.

I was forty before I realized that everybody didn't think just like I did. I didn't do very well in school and had come to believe I was stupid and would never amount to anything, but I didn't realize that all of the students who did better in school than me couldn't do three-dimensional imaging in their heads. I could do this with such detail that I could rotate an image, make changes, and see a working model that I could test by rerunning the model with the changes. That's the kind of ability I put to use designing and patenting gun parts. But I didn't realize I had a gift that was unrecognizable to my teachers because they judged my abilities on their own, and theirs were limited to their success in the educational system. It took me decades to discover how I was different and how I could use my natural abilities to my best advantage. Just imagine if the teachers in schools taught to all eight of Gardner's intelligences and all children could excel to the limits of their own incompetence. We should not be comfortable with the standardization of the two things that most affect what we believe and how we think: science and education.

I now speak often to students and educators about learning differences in education. Once, after delivering a talk called "Bears, Dyslexia and Education: The Price of Being Different" to a group of teachers at a special education center, one of the center employees came up to me and shared that he was dyslexic, but was lucky because the founder of the center had recognized his skills and had given him a professional job. He then asked me if I could spot people like us when I met them. I told him I could and that it took less than a minute. He pointed to the group of teachers and said, "They can't."

I find solace in the realization that one of the most corrupting aspects of human behavior is an instinct. If we can accept this then we can rise above the fray, using our minds and our ability to communicate to solve some of the many problems brought on when we as humans became successful enough to no longer be limited by nature and created a sustained surplus of food and other resources.

Modern humans have operated on a social model that evolved in the earliest of times and that evolved with only the information that was available at the time. In essence, when we learned to write the first thing we did was to write our own résumés. It would be hard to imagine that we could do this any other way. To this day we use this résumé we call humanity as the basis for the analysis of most of our actions. In the case of warfare or conflict, we seek to control weaponry as a means of ending war. At no time in our history have we looked toward the natural world, especially toward the social behavior of animals, to understand that we are a species in evolutionary transition and that we are out of control. We have focused on separating ourselves from animals, rather than looking at them as models that have been molded and shaped by five hundred million years of bit-by-bit development. Until we are able to get over ourselves, we will proceed down the path of a technology-driven arms race that may in the end be just that: the end. Understanding the common denominator of human behavior could, in the end, save us from our own technology.

To think this way requires hope. It also requires a firm understanding of what has shaped and continues, at some basic level, to motivate us to do what we do.

And that comes back, again, to sorting out information and making decisions. One of the huge costs of human success has been the loss of the ability to accurately collect and analyze information. Bears, as we've discussed, don't experience this same problem. If a rule has been broken, they know who broke it and have an honest and stable system of judgment and punishment with which to respond. They make decisions about where to find food, build dens, and form alliances based on concrete, firsthand information. A poor decision for Squirty means an immediate loss of fitness, which would directly affect her ability to promote her own genes, not to mention risk her own immediate survival and the survival of her offspring. Bears don't need to sort out good information from bad information. Yet sorting out good information from bad information is a job that consumes all humans throughout our lives.

Every day as we make hundreds of decisions, we hope we have made the right ones. We live in a world saturated with information. Information that reflects only what we think we know, what we want others to think we know, information with bias, information that seeks to influence us, information meant to deceive. Some of this information is correct and some is incorrect. We manipulate this information, we twist it to suit our purpose, we write it down for the record, we run it through filters, and we adhere to strict methodology and try to standardize it if we can. We promote, study, revere, and sometimes revise our own history. At the same time, we lack the ability to recognize good information when it smacks us in the face. At our very best we make decisions with the only information we have at the time; often we don't experience the unintended consequences of our decisions until many years later.

Arms control may be a means of preventing war, but warfare is a manifestation by humanity of the fairness, judgment, and punishment that works so well in Squirty's world. As humans, we practice going to war nearly every day of our lives, and we seem to relish it. We take sides on competing ideologies, on politics, on sports, on education. There is hardly any issue, from any part of our lives, for which no one can find an opposite side to be on. We can be passionate and even volatile with our emotions. As our populations grew, so did the size of our coalitions. These coalitions can act as a single individual in Squirty's world. Collective ideologies are formed with the effects of contagion, imitation, and mutual interest. Collective judgments act on a collective moral compass that often results in collective action. All of this can and has led us to war.

There is a huge advantage in knowing that our us-versus-them behaviors are instinct gone wild. We really can't help ourselves. When I listen to the world news, I see whole countries acting like schoolchildren, playing the elaborate game of emotional chess. My point is that if we recognize the fact that this is how we act and it is how we have acted since the beginning of time, then it should be possible to look beyond the day-to-day battle and find the root causes of kinds of problems that often lead us to war. If we can step out of the fray and observe, it is possible to find the crux of any problem. Unfortunately, humanity hasn't figured that out yet; there are far too many players, and not enough observers.

There is great irony in the fact that the only reason I was able to develop this hypothesis about black bears as a model of early social behavior, and create my own method of art and science, is that due to a poor ability to read and a high IQ, I was able to slip out of the net of confusion that is standardized education and science. I still live in a world much like Squirty's, where my best source of information is what I glean from observing and experiencing the world as it exists. The second irony is that my voice may never be heard because modern science has lost the software to open this file. Somewhere in our past we began relying on the writings of others and lost the ability to use the high-quality information our brains were designed to process.

We also lost sight of how to cater to our own needs while still being part of a broad, just social network. The independent spirit that is still revered in this country is just that–a spirit. Independence is a state where cooperation is no longer required; in other words, it's a state achieved when one can afford a solitary existence. While Squirty is often by herself, her survival depends on access to resources in areas occupied by other bears. Thus she has to be careful how she treats strangers whose favor may at some time be needed. There will be many times in her life where her fitness will be at the mercy of these strangers.

We all at some point wish to be free of social pressure. People get upset when someone noses into their business, passes an ordinance or law that intrudes on their lifestyle, or forces them to pay taxes when they don't agree with how the money is spent. In fact, there is little we can do without raising the concern of others. Why doesn't everybody just mind his or her own business? The answer to that lies again in Squirty's social behavior. It is an answer older than the dawn of man. All of these inconveniences are the price we pay for receiving shared resources. Receiving shared resources or benefiting from cooperation is the reason why we are nice to one another: We can't afford not to be.

There are some who feel they are above the law, others who feel they only need to comply with the law, and more still who are guided by religion, but in the end it will be the forces of nature that will define our boundaries. The simple reality is that we can run from the people we cooperate with, but we cannot hide. We are destined for greater and greater social control and a corresponding loss of independence as our populations continue to experience growth. It matters little if the population at hand is a local population or one on the world stage: Growth in population means our freedoms will suffer as the dependence on one another grows. It is interesting that we are born with an independent spirit and that each of us has to learn these realities.

Evidence of this trend exists throughout the United States and the world, both in our history and in modern times. We saw a sparsely populated America move from the lawlessness of the Old West to the more socially controlled small-town America to the highly controlled big city—all as population density increased and each person's number of cooperators multiplied.

Could humankind explain its entire transition from its early forms, when our individual existence was controlled with selective forces of nature, to the super-organism that we are today by analyzing the dynamic changes in social control that were required to manage the rapid growth and density of our human population? I believe we likely can.

Our history has recorded great success, great failure, and many unintended consequences for our actions. With our expanding wealth and prosperity, we have taken the responsibility for survival away from the individual and bet the future of the human race on the collective wisdom of humanity.

If we understand the conditions that have led to cooperation, altruism, and goodness and the corresponding problems of fairness, judgment, and punishment in the bear world, and if we understand that the good comes with the bad naturally, we will have a far easier time understanding why humans can be and have been so cruel. We only need to look at a stable society that has all the elements of our own to find a basis to understand our own society.

While the social behavior of the black bear makes a good model for an understanding of the origins of cooperation, altruism, morality, punishment, trade, and language, it is necessary to return to the toolmaking behavior of the great apes, primarily the chimpanzee, for a model of the origin of the technological advances that propelled us on the wild ride of humanity. It was environmental and evolutionary circumstance that combined the behaviors that exist independently in the chimpanzee and the black bear into one species, the human.

It should not be a surprise to anyone that a primitive model for human social behavior does not exist in primates because it has run its course in the human primate. The primates that exist today all have social behaviors that allow them to access their resources with stable groups (with the possible exception of the orangutan). Tool use has developed in primates to solve problems that they were not physiologically adapted to solve alone. But even though they use tools, their stable group social behavior may have limited any further evolution.

The black bear, according to my model, has the social behavior that might propel a chimpanzee-like creature into the unstable course that man has taken. But the species remains stable due to adequate physiological adaptations and communicative skills.

Let's consider three big ifs: If bears were not so physiologically fit for their environment; if they had the chimpanzee's need and ability to adapt by using tools; and if the effect of that adaptation caused a significant increase in overall fitness, then their social behavior might put them on a parallel course with humans. Tool use has allowed chimpanzees to adapt to new situations, but their social behavior limits the development and spread of cultural advances.

Current scientific theory suggests that man evolved from the social-group-forming chimpanzee. My model points to a potentially earlier branching of the evolutionary tree—suggesting that it is more likely that the chimpanzee evolved from a prehistoric human relative that already had a food-sharing social behavior similar to that of a black bear. Bears have the same independent spirit that exists in humans. They lack the grooming, begging, and groveling behavior that exists in group societies like that of chimps. Squirty's desire to dominate her resources provides a good explanation for why human societies have developed in many forms from primitive to modern. After all, it is still the distribution and demand for resources that dictates both animal and human social behavior.

So let's put it all together. The evolutionary success of humans depended on the need and ability to make tools, which led to an increase in overall fitness and a social behavior that was dynamic enough to take advantage of this increase in fitness. The result of this combination brought on an escalation of self-fulfilling change. Not only was there an escalation of problems to solve, both socially and technologically, but there would be an increasing number of people to solve them. As long as food and other vital resources are available to maintain these conditions, humans will continue to adapt and change socially at an evolutionary warp speed. Of course, the inverse is true as well. Meanwhile, the bears will be bears and the chimps will be chimps.

Applying an understanding of how this runaway freight train of humanity got its start also gives me clues to explain our recorded history—and if I'm lucky, perhaps it will reveal a course of events that could get this train under control.

In my mind, an understanding of what human instinct is will be the key to unraveling the cultural conflicts that dominate our world. A basic understanding of the built-in forces that have driven us from the beginning of time could allow us to step outside our present situation and understand it for what it truly is. If we know that the behavior that has driven every person on earth has a common denominator, then we could predict actions and reactions, allowing for reasonable solutions to religious and cultural problems. We might understand that we all got to where we are because that is simply the nature of our beast. We have likely been successfully out of control since the beginning of humanity. And we might use that knowledge to spin back into control.

In other words, when we understand our behavior in its simplest state, we will have a much better basis to understand human behavior in all its complexity—as it has evolved over time.

It is also helpful to understand that any organism will develop in ways that best serve its own interests and that of its relatives. That self-interest will be limited only by the availability of resources and the ability to reproduce. With a low population density and an abundance of low-quality food, an organism can be solitary, like the panda. With an abundance of high-quality resources that can be dominated, a society of close relatives can develop, as we see in social insects like leafcutter ants. In evolutionary terms, a species doesn't exhibit altruism until it has to—such as when it is faced with a highly irregular food supply and forced by circumstance to share short-term surpluses.

In my view, evolution is nothing more than the results of making simple decisions as correctly as possible. The decisions can be made either cognitively or physically. It is too easy to think, as humans, that we have created superior methods because of our success. In terms of evolution, the best decisions are made under the pressure of survival. Due to our surplus resources, we, as individuals, no longer have that pressure, but our species does.

Historically, humans have made decisions with only the information available at the time. Many of these decisions were made to better the human condition, or to maintain a surplus at the time the decision was made. But as we know, the effects of the unknown make it unlikely that decisions made in our past will serve the conditions of the future. To understand this, we have only to consider the consequences of our decisions to run the world on fossil fuels, or to use high-tech vessels to scour the seas to put more fish on our dinner plates, or to advance industry through the use of deadly toxins. The list could go on.

Our flawed decision-making process is complicated further once we throw in another primal behavioral trait that, in its simplest form, also leads to success: imitation. For wild animals, imitation is a powerful aspect of the learning process. Bears can learn from and mimic the experience of others to reliably and rapidly learn how to use their environment. The pressures of selection ensure the accuracy of the information that is passed. As humans, we still learn from imitation, but unfortunately as selection pressures have switched from the individual to the species, the information passed along in this way is no longer reliable. Yet instinctively we treat it as reliable, and we resist change.

Other aspects of human behavior also point to the fact that, on an individual level, we've lost touch with some of our earliest evolutionary prompts. It has been interesting to experience bears' awareness of emotional communication and their ability to use intentional communication to manipulate the behavior of others. People, on the other hand, often seem relatively unaware of their own emotional communication–despite the fact that it remains our most common mode of getting our point across, and we use it every day. Yet we are considered self-aware and by most accountings bears, and other nonhuman animals, are not. Yes, we can talk about ourselves, yet we have lost the ability, in many cases, to consciously respond and intentionally control our emotional responses to manage our own emotional situations.

There are endless examples of this lost instinct. A police officer can get drawn in emotionally by the individual he or she is trying to arrest and, as a result, act emotionally rather than cognitively. When we argue with our partners and spouses, the emotional battle draws out with little cognitive understanding or management. In politics, whole parties and even whole countries can get drawn into emotional situations with no true understanding of what they are being drawn into.

Why are we so vulnerable to being duped emotionally? It's because our emotional systems evolved, over the course of time, to function with high-quality firsthand information–quite unlike the kind of information we often encounter.

Books, articles, documentaries, and more all love to probe the question of how cooperation and altruism began in humans. It is–like the advent of language or our use of symbols–one of the things that perplexes and intrigues us. The answer, though, appears to be that cooperation and altruism didn't "begin" in humans at all. These traits existed, likely, long before the first hominin set foot on this earth, brought on by a simple extension of selfishness in a landscape rife with food surpluses here and shortages there. It was launched by a sentiment as simple as "This is mine, but I won't go after you if you take it because I have all I need." And as competition for resources increased, the message changed: "You can take my surplus, but on the condition that when I'm in your position you share with me."

Black bears can provide an excellent present-day model to explore the primal elements of this kind of social exchange. But even then mysteries will remain as we begin to probe the origins of the bear's behavior–something that also likely can trace its roots back millions of years.

Chapter Eight

From Black Bears to Giant Pandas

When I presented my first accounts of bear behavior, scientists didn't want to touch my findings with a ten-foot pole. They couldn't get past my controversial methods. The packaging wasn't just right, either; my observations weren't published in a peer-reviewed scientific journal, and nobody else was doing this kind of research on bears of any kind, so it was hard to find a scientist qualified to discuss it. There are lots of people doing this kind of work on apes and monkeys, but they can't speak to bear behavior. And there are plenty of bear experts out there, too, but most of them are wildlife biologists who have never taken bear behavior very seriously, although I know many of them have some very strong opinions. Their specialties include population dynamics, management, and ecology. Subjects of behavior including bear nuisance problems and bear–human interactions have typically been treated reactively and with very human techniques–like cracker shells and bear spray.

Some prominent scientists sought me out and were intrigued by my findings, and I was invited to speak at scientific forums on occasion. But by and large, I have spent most of my bear-studying years well outside the academic or professional fraternity. In many ways that's been good for me: I haven't had to follow unwritten rules that can cloud research, or move my work in certain directions to attract funding or stay out of another scientist's research realm. I have just done what I've done, methodically, and with observation techniques that seemed in concert with the writings of Nobel Prize–winning ethologists like Konrad Lorenz and Niko Tinbergen.

Along the way, I have unwittingly broken some of these unwritten rules. But apparently my biggest "transgression" has been my decision to raise orphan cubs as a surrogate mother. It made perfect sense to me. First, they were orphaned, and following the then-common practice of releasing them back into the wild at five months seemed like a death sentence for them. Second, how would I or anyone else, for that matter, know what orphan cubs' needs really were if their development and juvenile behavior weren't documented? And how, I thought, could anyone ever understand the behavior of an adult animal if he or she didn't work with the animal as a juvenile? When I started out, it never occurred to me that what I was doing would stir such wrath.

It's not that the world of credentialed science was unknown to me. My family life was steeped in it. My father, Lawrence Kilham, was a virologist at the National Institutes of Health (NIH) and Dartmouth Medical School; my late brother Peter was a professor of limnology at the University of Michigan; my mother, Jane Kaufolz Kilham, and my brother Michael were physicians; and my sister, Phoebe, has her doctorate in tropical soils. But it was my father's work as an amateur ornithologist that influenced how I approached my own work, and perhaps even prepared me for that work being challenged by the scientific community.

He started documenting the behavior of birds when he was about forty, though he had been fascinated with nature and particularly birds since he was a child. As a Harvard undergraduate he became friends with Ludlow Griscom, often considered the father of modern birding, who recommended him as a member of the Nuttall Ornithological Club. During World War II, he and my mother each joined different branches of the Red Cross and went to England as medical doctors. Father ended up crossing Europe with Patton's Third Army, attached to a field hospital. During lulls in the fighting, he continued to watch birds and keep journals of his sightings. But it wasn't until he moved to Maryland to work at the NIH that his interest in studying the behavior and life histories of wildlife, and particularly birds, truly manifested itself.

As he wrote in his book, *On Watching Birds*, "My start as a behavior watcher began in no dramatic way. I knew I wanted to learn more from the wonderful countryside where we lived in Bethesda, and I knew I must originate some systematic way of doing so. Descartes advised anyone undertaking a new enterprise to begin with what is simplest and easiest, and that is what I did in the early 1950s. Whenever I took a walk or went anywhere, I began observing with the first birds I met."

He went on to document the life histories of the woodpeckers of the eastern United States and published more than 125 journal articles on bird behavior. Once, though, he had the unpleasant experience of publishing a paper on the endangered red-cockaded woodpecker, and swore he would never study an endangered bird again. He was attacked by professional ornithologists who apparently felt his lack of ornithological credentials called his work into question, or they saw him as competition.

He did not get the same response when publishing his studies of common birds, though, and soon started studying crows and ravens. He spent more than eight thousand hours documenting how they foraged, stored food, bred, vocalized, and carried out a host of other behaviors, which he documented in his book *The American Crow and the Common Raven*. Late in his life, he received a call from the US Fish and Wildlife Service, which told him that his book on crows was instrumental in their efforts to save the Hawaiian crow—called the alala—whose numbers were so low they could not risk interfering with the remaining birds to collect the kind of information he had documented. By the time he died, the alala was extinct in the wild, and its future was in the hands of captive-breeding programs.

He maintained that it was far more important to study a common species, because at any time they or one of their related species could become endangered.

I didn't know then how pertinent that advice would become. I didn't set out to develop methods that could be used to save or reintroduce threatened or endangered bears. I did what I did for my own reasons. I had grown up assisting my father with his bird work and had become good at making detailed observations. My less-than-perfect performance in school and the judgments that came with that had long-term effects on my confidence. I had learned from my experience as a gun designer that using my own methods, I could be successful. I had been interested in studying animal behavior in a manner similar to my father's, but those weren't the methods taught in school. So when I started working with the first set of bear cubs, I was determined to simply learn from watching and use my own wits to understand the bear's behavior.

In 1996, though, the acclaimed field biologist George Schaller called. He was going to a workshop with the World Wildlife Fund and Chinese officials to discuss the reintroduction of giant pandas into the wild, and he wanted me to write a short paper detailing the methods I had used to rehabilitate and reintroduce black bear cubs to the wild–and also to suggest how they might be used to reintroduce pandas. Many captive-breeding programs faltered when animals reared in captivity had to make a life in the wild, and scientists were searching for successful models before they released the pandas back into China's bamboo forests.

So I wrote an account of walking cubs to give them an opportunity to learn about and have confidence in the natural environment. I explained the importance of eventually placing the cubs in remote cages to protect them while they interacted with older resident bears in a wild setting, while still having refuge and supplementary food. I gave my insights based on my experience with successfully releasing adult ravens that had been in captivity. After the workshop concluded, George sent me an abstract he wrote–"Giant Panda Biology and Its Relevance to Reintroduction Efforts," which outlined the panda's habitat considerations, food source (bamboo), and social life. In his recommendations for important elements to consider when reintroducing giant pandas, he incorporated my suggestions of walking cubs, and the use of remote enclosures to provide secure releases of subadult and adult pandas. In a handwritten note on top of the abstract, George wrote, "I just returned from this workshop. I'm not sure of the next step. Thanks for your help."

Many years would pass before my first trip to China to present my work at the Chengdu Research Base of Giant Panda Breeding, informally known as the Chengdu Panda Base.

But before that journey I would hear from other bear researchers around the world working with threatened populations.

In 2007 I was invited to attend and present at the International Workshop on Bear Species Rehabilitation, Release and Monitoring, held in Bubonitsy, a town in Russia's Tver Region, and sponsored by the International Fund for Animal Welfare. There, bear rehabilitators from around the world, working with all eight species of bears, gathered to share what they'd learned about reintroducing bears, young and old, into the wild. The goal was to come to a consensus on bear rehabilitation and release methods–something that was generating increasing interest. In fact, the polar bear was the only species that was not being rehabilitated and released, and that was because of the difficulties in dealing with the Arctic environment. The time was right to establish criteria for accepting orphan bear cubs into rehabilitation programs, develop guidelines for their care and rehabilitation, create frameworks for deciding whether or not a bear is suitable for release, and find the best protocols for the release itself, post-release monitoring, and public outreach.

We gathered in Moscow and traveled by bus to Bubonitsy along the major truck road from Moscow to St. Petersburg. The road was rough; many times the traffic slowed to a crawl due to the bumps and potholes. I could see from looking at the dirt shoulders of the road that they were well traveled, often smoother than the road itself. We stopped at a newly built tourist area; it was large and nicely built, but there were no tourists and the buildings appeared empty. There were restrooms and fuel, but the whole facility was a contrast with the road itself, which was lined with abandoned agricultural fields and traveled by vintage tractors, circa 1940, and old military trucks carrying logs.

As we entered the town, the houses were old and rough, with electricity coming in on hot wires and glass insulators. It was as I imagined much of rural New England in the 1930s. Every house had a vegetable garden, and the whole place had the feel of a rural peasant village. But it wasn't–a fact belied by an abandoned enclosure for captive wolves. It was an enclave of Russian scientists.

Bubonitsy was the home of Valentin and Sergey Pazhetnov, who had been rehabilitating Russian brown bears since 1982 in a facility near their home. A large log conference building had been constructed especially for the conference. Later, it was to be used as an interpretation and education center for the Russian Orphan Bear Cub Rehabilitation Project.

This was my first trip away from North America, and we were warned that the facilities would be rustic. We were given options of sleeping in tents or bunk buildings, or staying with Russian families. I opted to stay with a Russian family. Because of the language difference I was not always sure what was up, but I followed along as I was taken to the home I was to stay. Here I met a man who was talking about polar bears and Wrangel Island. In *Among the Bears*, I had written about Nikita Ovsyanikov and his book *Polar Bears: Living with the White Bear.* He was a man after my own heart who studied polar bears by observing their behavior. So I asked this gentleman, "Do you know Nikita Ovsyanikov?"

He replied, "That is me, Nikita Ovsyanikov."

What were the chances that on my first trip to Russia I would be able to spend a week with the man I truly admired? As deputy director of Russia's polar bear reserve on Wrangel Island, Nikita is responsible for protecting polar bear habitat, and their populations, in and near the Wrangel Island Reserve. We had long discussions about methods of research, and I found solace in the fact that he had used some of the same techniques throughout his career–first with arctic foxes, then with polar bears. When I told him that my research methods weren't widely accepted by modern scientists in the United States, he responded that he thought such methods were the only effective way to conduct research. It was great to hear positive feedback from a like-minded scientist.

Many of our experiences were similar. Nikita invited me to view a documentary he shot that was produced by the Discovery Channel. I say he shot the film because in the first effort to make the film, two filmmakers from Discovery traveled with him to Wrangel Island. "The first day there were over a hundred polar bears, the next day half that number, and the third day there were none," he recalled. So the filmmakers went home. Nikita returned the next season with his own video camera and shot all of the footage, and the Discovery Channel produced and edited the film.

He had warned them that this would happen, and I often experience the same thing with Squirty and strangers. Bears build trust with individuals, and it has to be earned.

There was only one television with a VCR in the community, and Nikita arranged for a time we could use it. Several of us walked to the house after midnight to view the film. As I watched the remarkable footage, I found I was able to predict the behavior of the polar bears, as much of their social behavior was the same as I had observed in the black bears. When a subadult polar bear approached an older bear with an open-mouth display to solicit food, he got it–just as a black bear would using the same behavior.

There are common behaviors in the bear world, just as there are common obstacles in the world of bear experts and rehabilitators. Some of us have more obstacles than others. Nikita, for instance, works with a population of bears severely threatened by climate change, while I work with a population of black bears whose numbers are limited most by human tolerance, not suitable habitat. We all have our different methods, as was apparent at the start of the Russian conference, in a room full of bear experts each trying to promote his or her own methods. But there were signs that the bear-rehabilitation world can get past such rivalries and come together over what works and what does not. By the time we were sitting by a lake and listening to traditional Russian ballads as the sun set on the last night of the conference, tensions had surely faded.

While my work revolves around black bears, I do occasionally dip into the worlds of other bears to solve riddles in my own work. In fact, my interest in pandas began with such a riddle. I had gone to Harvard in 1997 to give my presentation to Richard Wrangham's graduate students and colleagues; while there I stopped in at Harvard's Museum of Comparative Zoology and asked to see the skull collection. The museum's curatorial associate, Judy Chupasko, led me to a vast collection where I was able to locate and inspect skulls from a giant panda, other bear species, and a wide variety of other mammals. I was looking for clues that might help reveal if other mammals had the same variation on the vomeronasal system I had discovered in the black bear. I could tell from quick visual inspection whether or not each skull had a pocket in the vomer for the Kilham organ.

I was able to find the pocket in the skulls of other bears, including the spectacled bear, which is the last of the short-faced bears and an earlier branch of the bear family, but in the giant panda and other mammal species, it was missing. I had suspected that the Kilham organ would be missing in panda skulls as they were specialists who ate primarily bamboo, while other bears were generalists eating a wide variety of food.

The fact that it was indeed missing suggested that pandas would behave differently to the scents that excited black bears. So in 2006, with the help of Kati Loeffler, a wildlife veterinarian who had worked at the National Zoo in Washington, DC, we made an application to the zoo to do a few experiments on the giant panda using scents that had excited black bears. The results would give me insights into the marking behavior of both black bears and pandas. It took more than a year and a half and a huge number of emails to finally gain access to do my experiments. I wanted to record with video the first-time reactions to the scent sample I brought down, which were spruce and red pine pitch and citronella bug spray. My experiments were simple; they had to be, as I lacked any funding. I planned to just offer three scents–spruce pitch, red pine pitch, and citronella–and record the pandas' reactions on video.

The keeper at the panda enclosure led me to my work area behind the public displays, near the banks of video cameras that monitored the pandas and the food preparation areas. I was able to access the pandas through stainless-steel bars, offering them opened ziplock bags containing the pitch and spraying the citronella from a pump bottle onto a piece of cardboard that I slid under the bars. As I expected, the pandas had little interest in and showed no excited response to the smells.

While I was there, the keepers invited me to try my scents on the spectacled and sloth bears. The spectacled bears were definitely second-class citizens at the zoo: Their back area was dark and dimly lit, and my access to the bears was poor at best. I was unable to see well enough to observe a response. But the sloth bears had new and well-lit enclosures where I could easily observe and film their responses, and I arrived when the bears were inside them. There were three cells with four bears, a mother and five-year-old son, an adult male, and a four-year-old female who had recently arrived from another zoo to mate with the adult male.

I approached the enclosure of the four-year-old female first and opened my bag of spruce pitch. To my surprise, she blew it out of my hand. This was new to me. Sloth bears feed on termites and have powerful abilities to suck and blow to extract termites from their mud mounds. They had a demonstration for the public where Merlin, the large male, would suck up mealworms from the hand of a keeper standing with the audience through a ten-foot plastic tube. Then I sprayed some citronella on a piece of cardboard and slid it into her cage. She responded by running across the room at me and jumping onto the bars with a big smile on her face; then she played with all of her enrichment toys. I noticed that suddenly there was an audience of keepers and I said, "Doesn't she do this all the time?"

"No," they replied, wanting to know what I had done to get her so excited.

I tried the citronella experiment on the female with the five-year-old son next, and he responded by rapidly batting the cardboard with the smell out of his cage. Then it was the large male's turn. He came over and sniffed where I had held the cardboard with my finger, and then he false-charged me.

The experiments were a success, except for those with the spectacled bear. The panda showed little or no response as it lacked the Kilham organ, and the sloth bears had a response that was very similar to the black bears' own: The female got excited, the subadult male did not want to get involved, and the big male made a threatening move, seeing me as competition for a mate. I had found that aromatic scents elicited an excited response from females and suspected that it was because males bit trees to release aromatic sap to advertise their availability during the mating season. These were the smells of courtship.

It wasn't until the spring of 2008, though, that I would get to see pandas on their own turf. That is when I was contacted by James R. Spotila, the Betz Chair Professor of Environmental Science at Drexel University, about going to Chengdu Panda Base to present my work to the Chinese scientists who were moving forward with reintroducing giant pandas there. I had met Jim when speaking at the Monell Chemical Senses Center about the Kilham organ. He had been given *Among the Bears* by my sister-in-law Susan Kilham, also a professor at Drexel, and had apparently liked it enough to come and hear the talk. When the trip to China was being planned by the Global Cause Foundation–run by Jim's brother John Spotila and his wife Sheri Yi–Jim felt it was important that I go along.

Global Cause had been providing scientific support to the panda base, then home to 115 giant pandas and the center for their captive breeding. In practical terms, this was a treasure trove of insights into the panda world. There are only 1,600 giant pandas left in the wild and 300 in captivity. Despite some effort, the Chinese have not yet been successful in releasing captive-born giant pandas back into the wild. And they wanted to be as skillful as possible in their next attempts.

Our delegation arrived in Beijing, and Sheri arranged a mini immersion into China's modern and ancient culture. We were taught never to drink at dinner without proposing a toast along with a host of other customs we'd need to know as we spent time dining formally with our Chinese hosts. (The Chinese spend time getting to know and befriending new colleagues before discussing business–quite different from the Western approach.) We visited Beijing, Xian, Shanghai, and Chengdu, seeing the Forbidden City, Tiananmen Square, the clay soldiers, museums, and universities. We got a sense of the country's stark contrasts as we saw modern skyscrapers reach upward and bullet trains swoosh by while workers swept the streets with straw brooms and carried heavy loads on bicycles. Finally, we were ready for the Chengdu Panda Base. Here we were greeted by Hou Rong, the director of science and research at the center. The base had lovely manicured gardens, tour buses, tourists, enclosures with both giant pandas and red pandas, and a large pond with black swans and thousands of multicolored koi that the tourists could feed. We toured the base, had pictures taken with panda cubs in our laps, had meetings, and dined on extravagant meals. On the second day we gave our presentations. I was the last to speak, and unlike the presentations of my colleagues on the trip who were all professional scientists, I gave the public presentation that I had been routinely giving at home depicting what I had learned from walking with bears. At the end of my presentation, Hou Rong came up to me and said, "We noticed that your presentation was very different from the other presentations." Then she added, "You make us think."

In the morning we were off to see giant panda habitat. On our way to the mountains, we drove through the rubble of the May 12 earthquake that devastated much of Sichuan province. Our first stop was at the Department of Forestry office, where I was informed I was to give my presentation again. I wasn't prepared and didn't have my computer with me, but that wasn't a problem–my presentation was sent up from the base via email. The forestry officials presented first with a program showing the effects of the earthquake on giant panda habitat. Much of the steep terrain in the panda reserves had slid, and mountain villages were either destroyed or badly damaged. It wasn't clear how many giant pandas were lost. The Wolong Giant Panda Breeding Center had been destroyed.

After the presentations, we made our way in the bus to see the giant panda habitat. We followed a long and winding road up through one destroyed village after another until the road ran out. On foot we hiked to see the mixed bamboo, softwood, and hardwood forest that the giant panda called home. Unfortunately, there were no pandas occupying this habitat.

Before we left, the Chinese revealed their plans to build a giant panda reintroduction and education center, Panda Valley. But it would be another four years before I returned to China to see the progress.

This time Jim, John, and I flew to Beijing, where Sheri was waiting for us at the airport and Kris Cena, a professor and specialist in metabolic rates in animals, arrived from Poland to join us. We all made our way to Chengdu, where we were joined by Yuxaing Fei, Jim's graduate student who was doing his PhD work on metabolic rates of giant and red pandas. We were at the panda base for ten days, and while Jim, Kris, and

Yuxaing worked trying to calibrate instruments and finish the metabolic chambers they would use for their experiments there, I concentrated on observing captive giant pandas.

Panda Valley was near completion. It had taken $4.75 million, but they had successfully fenced in 320 acres that was being reverted to panda habitat. I had hoped that plans for the reintroduction would be well under way, but discussion about the best release methods was still ongoing. It is risky in more ways than one to work with endangered animals. You never want to jeopardize any animal, but when your actions with just a few animals could affect such a percent of the population, you need to take extra precautions. Still, I have always found that the best way to be successful in a project is to start it, and the methods of reintroduction used with black bears ensured minimal risk of failure as long as the experience and knowledge gained as the project proceeded were continually applied.

I also had a chance to catch up with some of the scientists who were working out plans to eventually release bears. Hou Rong was still eager to learn more about my methods of walking with black bear cubs to introduce them to the wild and had lots of questions about how many we worked with, what our enclosures were like, and other specifics. She had not lost the enthusiasm she showed about using the black-bear model when we first met, and I held out hope that plans would move forward.

While in China this time, though, I wanted to see actual giant panda habitat, so I asked to see giant panda scat in the wild. So Wen Ping Zhang, one of the researchers at the base accompanied me to Longxi-Hongkou National Nature Reserve, at the headwaters of the Minjiang River and punctuated by Guangguang Mountain, which stands 15,032 feet tall. There are an estimated twenty pandas in the reserve and many other species of rare plants and animals. The damage of the 2008 earthquake was still evident as we wound our way up a steep road with dozens of switchbacks to the reserve, which had been quite close to the epicenter. The habitat was steep with ridges forming what appeared to be a horseshoe shape around a flat river valley–a field of stones, with water no longer flowing on the surface. Landslides were grown over with vegetation, and the sounds of long-tailed macaques filtered through the thick vegetation.

Tao Shang, the director of the reserve, led us in our quest to locate scat, guiding us along a narrow foot trail before we made our way to a thick stand of bamboo. I noticed a pile of bamboo husks and some shoots cut with a sharp clean cut and wondered aloud if it was a sign of bamboo rats. No, they said; it was a sign of people. We were in the core area of the preserve, which was supposed to be free of people, but we saw many more signs of people and heard them in the bush. These were poachers illegally harvesting bamboo and competing directly with the pandas for food. Tao and Wen yelled out at them to leave, but didn't expect success in tracking them down. Poachers know the bush well, and are expert hiders. It wasn't long before Tao found giant panda scat as well as signs of pandas feeding on bamboo–something they do by biting into the shoots with their front teeth, leaving behind an irregular cut mark that differed significantly from the clean knife cut left by humans.

We worked our way back to the trail and continued on into the reserve, eating strawberries and observing ungulate scats, butterflies, and insects. As we crossed a boulder field that had once been a river, we spotted two bamboo poachers fleeing into the thick vegetation. Tao Shang pointed up into one valley and said they had found an emaciated panda there just two weeks before and had taken it to the panda base to be cared for. Later I heard it was missing a front paw–a sign it had been caught in a snare. It was apparent that controlling people would be the single biggest challenge in the reintroduction of pandas into the wild, and perhaps the major threat to their survival. Despite strict laws on poaching pandas, enforcement was a problem.

I had gotten one step closer to wild pandas at Hongkou and spent the ride home yearning for a chance to return to the remote area and experience the giant panda in the wild.

By the time we arrived back in the States, despite the meetings with the panda base staff and the efforts that had been put into Panda Valley, I still felt that I had not connected. The Chinese lacked the essential experience of rehabilitating or reintroducing other animals to the wild. The whole notion was new to them, and it was difficult for me to convey to them with words and pictures alone. They would just be mixed with the rest of the advice they were receiving from many other sources. Fortunately Sheri had a plan. She had arranged for Hou Rong and two other scientists from the base to study English at Drexel University for two months; during that time they would visit us in New Hampshire for firsthand experience with our methods of raising black bear cubs.

So Jim Spotila drove our Chinese guests seven hours from Philadelphia to Lyme; John and Sheri were to arrive the next day. Debbie had not accompanied me to China, so having everyone stay at the house provided a great opportunity for all to meet. Phoebe and I took the visiting panda experts in to feed the cubs every day during their stay. One day after feeding we opened the door to the cage, and six cubs ambled out to follow us for a walk into the forest. The cubs paid no attention to us as they explored, found a hornet's nest that had been taken out by a wild bear, climbed trees, and sampled vegetation. Rong was amazed that they didn't just run off. I led the group to a red pine mark tree that I knew would have wild bear scent on it. The cubs all were excited as they climbed and checked out the bear scent. When they were finished, they all did a stiff-legged walk while marking with urine. Rong exclaimed, "They can communicate with the wild bears without even meeting them."

We made our way back to the cage, and all six bears followed us in. Rong was clearly excited; indeed, she and the others had not understood what I was trying to communicate in China. It was the kind of thing that needed to be witnessed to be duplicated. They could see from even a short walk that they could control the cubs, learn from the cubs, and give the cubs an opportunity to learn in a complex natural environment. They understood now why I had emphasized the importance of walking cubs where there were wild bears, since it is just as critical for the cubs to learn about their social environment as it is to learn how to find food. Our facility–an eight-acre forested area fenced with high-tensile electric fencing with a plastic deer fencing material clipped to it–is low-cost and a perfect model of what could be used for panda cubs. The most important aspect is that it is in wild bear habitat with wild bears routinely visiting the cage that houses the rehabilitating cubs.

That evening, Rong and her colleagues came with me as I met with Squirty and the other bears at my research station on my Lambert lot. The usual females and their cubs showed up, a total of fifteen bears. My Chinese friends were excited to see wild bears, and to their surprise they didn't feel threatened by them at all. At dinner that evening Rong said in just a few days we had turned their thoughts about the reintroduction of pandas upside down.

The interaction with the Chinese would, in its own way, turn my world upside down, too. Jim Spotila suggested it was time for me to get my PhD, offering to advise me at Drexel while I did my academic work and translated my years of recorded data into a doctoral thesis. My years as an academic outsider are about to end.

And so I have begun yet another interesting journey, all thanks to bears.

Appendix

The Human–Bear Conflict:

How to Understand Black Bear Behavior and Avoid Problems

Up to 900,000 black bears live in North America, and in many regions, like my own, they live in close proximity to humans. Or perhaps it would be more accurate to say that in many regions humans live in close proximity to bears–and that we are moving deeper and deeper into their habitat all the time. So, it's not surprising that bears and people meet up unexpectedly, and frequently. But of the millions of interactions between bears and people every year, very few result in human deaths. In fact, when scientists Stephen Herrero and Andrew Higgins studied all the fatal wild black bear attacks in Canada and the United States from 1900 to 2009, they found only sixty-three reported deaths.

Bears, on the other hand, have not been so lucky. Many are shot, either as a fear-based first resort or after other techniques have failed to deter what we've come to call "nuisance" bears. These are the bears that wander into backyards, campgrounds, landfills, or other places where food is often lying around. People can and sometimes do get injured by these nuisance bears, but even these incidents could be mitigated by understanding how to read and understand bear behavior. Not only would this knowledge help officials deal more effectively and humanely with nuisance bears, but it would also help individuals who find themselves in bear–human encounters.

In short, the solution to the nuisance bear problem is not so much about managing bears; it's about managing people.

Bears and Food: What You Need to Know

The most dominant drive in a bear's life is the drive to eat. Bears eat to store fat, which they need to hibernate, reproduce, grow, and endure food shortages. In fact, a bear needs to increase its body weight by about 30 percent to survive the winter, and a bred sow will need to increase her body weight by at least 50 percent to give birth, nurse, survive the winter, and feed her cubs in the early spring. Like all animals, they also prioritize the food they eat according to quality and quantity, as well as the risk involved obtaining it.

When natural food sources are in short supply, birdfeeders, garbage cans, and other easily available outdoor food become the highest-quality bear habitat around. This explains why bears exhibit more nuisance behavior in months or years of natural food shortages than they do when abundant natural foods are available to them. Bird food and animal feed, for instance, have two to three times the caloric value per unit of any natural foods and are available in high volumes in many residential backyards. Though bears are attracted to even small amounts of bird food in a feeder, it is not unusual to find 30 pounds of black-oil sunflower seeds in multiple feeders outside a home. Dumpsters often have huge volumes of food–as much as 40 to 100 pounds a week.

So, the best way to end what we consider the nuisance–bear behavior is to just stop inviting bears to dinner. If the food sources in problem residential areas are reduced to a minimum, these areas will no longer be worth the risk to the bear and the problems will cease. How to do this?

- Remove bird feeders, and any other food placed outside to attract wildlife.
- Don't feed pets outside.
- Keep any livestock feed indoors.
- Don't put kitchen scraps in your garbage can. Composting your kitchen scraps in a smell-proof way is as good for the environment as it is for avoiding bear encounters. Try a bear-proof composting container, or an indoor vermiculture bin (in which worms help digest the waste). Or, if you're using an open compost pit outside, layer fresh waste underneath material that is already decomposed, or add a layer of lime, wood ash, or sawdust to mask the odor that can draw a bear's interest.
- If you cannot compost, then secure your garbage can in an indoor area, such as a garage, or freeze your garbage until it's time for disposal.
- Use bear-resistant food containers while camping, never keep food of any kind in your tent, and follow local guidelines for cooking or disposing of anything that smells of food, even the water you've used to wash your pots, pans, and dishes.
- Clean outdoor grills, barbeque pits, and coolers after use to remove odors.
- Remember, the secret to controlling bears is controlling smell.

It is important, too, to understand that when a bear seeks food from a human source, it has no idea it is breaking any rules. In fact, in the bear's world there is an expectation that surplus food will be shared unless it is defended, and a corresponding expectation of reciprocity. As a result, the sharing of food carries a message of friendship and trust–just as it does in our own culture. We invite friends over for dinner; we attend community dinners and even state dinners. We don't invite people over for dinner then call the police to accuse them of stealing food. And we need to stop tempting bears with outdoor food then trapping, relocating, or shooting them when they show up and keep coming back for more.

For one thing, removing a problem bear doesn't solve the problem. It just creates an opportunity for the next bear to occupy the newly available habitat niche. Bears mark their food sources, whether a berry patch or a bird feeder, leaving behind scent that identifies them, stakes their claim, and indicates when they were there. They also leave olfactory trails that other bears can follow to find the food. So, if one bear knows about your birdfeeder, you can bet others do as well.

Bears also actively share their surplus resources, whether found in the wild or in a backyard. They have friendly relations with other bears that share food with them and combative relations with ones that don't. People don't seem to understand that when they feed a bear–intentionally or unintentionally–they are entering into a social contract with them. Once you start feeding bears, they expect the food to keep coming, and when you miss it for even one night, the bears might respond by damaging property as a means of punishment. This is because bears have rules for sharing surplus food, and punishments for those who break the rules. This simple reality can explain a lot about why bears become increasingly problematic when previously available food is removed, or made more difficult to access.

It's no surprise, then, that when people do start feeding bears, it ends badly. They get into a situation that they can't stop by themselves. There are, though, nonlethal measures that can be used to resolve the issue.

Aversive Conditioning

Throughout this book I've described numerous examples of how bears enforce their rules. When Yoda's two-year-old brother, Houdini, was hanging around too long in Squirty's domain, she enforced her rules by ripping a three-by-five-inch piece of skin from his ankle, causing him to flee the area. On a separate occasion, she chased him a mile and a half, treed him up a large white pine, followed that with a stiff-legged warning display, then marked her sunken footprints with urine to let him know who it was that wanted him out. That done, she lay down and slept for a few hours while Houdini remained captive up the tree. Later, he left the area.

It is possible to use these same principles of aversive conditioning on bears in residential areas. But it is important to remember that bears have been managing their social food sharing with punishment for over six million years, while humans have been practicing aversive conditioning with bears for less than twenty years, with little understanding of bear behavior. As a result, many of the methods used to punish bears have been anthropocentric–that is, they make a lot of sense to humans, but may not mean much to bears.

For instance, we associate gunshot with danger even as children; it's part of our culture. So, we fire warning shots over bears or use poppers and bangers to keep them away. But, there is no natural experience to which bears can relate these sudden loud sounds–unless the bear has already been trained to react to a gunshot. I was walking Yoda and Houdini as cubs up our driveway when my wife's car alarm went off. We thought the sound would terrify the cubs, but they remained calm. Conversely, when deep in the forest with bears, I have often seen just the snap of a twig put them on full alert.

We also try to find a magic scent that will repel bears. I have received a number of calls from people who have marked, or wish to mark, the boundaries of their property with urine in an effort to keep bears away. But human urine, because we don't use it to mark, carries no signature, and therefore can carry no message of dominance. With a combination of scent and aggression one might be able to connect a message to the scent, but this would work only on the bear or bears that experienced the aggression. It would have to be repeated over and over again with other bears who filled the vacated bear's niche.

Bears, on the other hand, can exact punishment one-on-one in a timely fashion, can identify cheaters beyond doubt (thanks to one of the most advanced olfactory systems in the animal world), and enforce rules far beyond the limits of human forensics. They exact only the amount of punishment required to offset the infraction. And that punishment may come in the form of a false charge, a negative vocalization, a range of modified bites, all-out attack, or even death. Bears have reasons or motives for all of their aggression. Dominant bears chase; subordinate bears run. Dominance belongs to the aggressor. It is also true that an individual bear may be dominant in some situations and submissive in others.

By studying their behavior, we can increase the effectiveness of our own attempts to modify bear behavior. In my experience, when bears enter our domain, or the domain of another bear, they expect aggression and intend to flee. When they experience nonaggression, they assume the food is surplus and free for the taking. But how can humans deter bears like other bears do?

It is known that punishment or classical conditioning needs to be applied within 1.2 seconds of the act. With this in mind, I have observed and documented how black bears communicate aggression to each other, and I have mimicked this behavior with backyard "nuisance" bears. I refer to the resulting technique–essentially a human-dominance method–as "working a bear" or "walking a bear out." I have used it successfully in many instances where I am called upon to respond to nuisance bears.

The most difficult part of this method is building up the courage to try it. But other first responders who attempt it will likely find the results come quickly and easily. The more experience you get, the more effective you become at making bears leave an area. First responders on bear calls usually carry a gun, and those trying this technique might still want to do so for added protection. Soon enough, though, if done correctly, you will discover that it is not needed.

I learned this method from Squirty, the bear I raised as a cub and follow in the wild. When she has young cubs to protect, she always lets me know when I should leave her area, and with one of her early litters of cubs I experienced one of her most forthright methods for persuading me to go. She locked her eyes on me with a hard stare while walking slowly and steadily towards me. That message was unforgettable.

So, when I mimic this technique on a nuisance-bear call, I lock eyes with the bear and walk slowly and steadily toward it. Locking your eyes on the bear identifies it as your subject, and walking toward it not only demonstrates intent. It is interpreted as a very aggressive and dominant action. Residential bears, knowing they are in my territory, will usually turn to flee on my first step. The message can be enhanced by walking stiffly and by pursuing the bear until it trees or snorts and runs off. The longer the pursuit, the more effective the action. Generally, it isn't necessary to pursue a bear for more than 100 to 200 yards before they snort and take off. Bears are persistent; they may try to sneak back in for the food. If they do, wait for the bear and walk after it again. They will get the message and will be reluctant to return.

A bear who trees is showing its submissiveness and is very likely young and inexperienced. In this situation, I usually use the "huh, huh, huh" vocalization, a chesty reverberation that has a negative connotation. Depending on the intensity, it can mean, *no, stop,* or *go away and don't come back*.

In effect, when you put this method into practice, you become the dominant animal. This is a role played routinely by police and conservation officers: they wear a uniform and present themselves as the dominant figure when dealing with "nuisance" humans.

The value of this human-dominance method for first responders is that they will feel more confident about working with bears; and they will be able to get the bears they can see to leave the area immediately and, sometimes, forever. Another advantage of this method is that it doesn't cost any additional money to apply. First responders feasibly can quickly and effectively get a bear to leave the area with only a working knowledge of behavior and what is routinely carried in a vehicle–like a bullhorn, which can be used to amplify the negative bear vocalization as part of the aversive conditioning process.

The other advantage is obvious: it saves bears' lives, and keeps them from being trapped and relocated.

Ultimately, though, the food attractants that caused the problems have to be addressed.

Diversionary Feeding

Bears have evolved to endure natural food shortages. Still, our forests, for a number of reasons, fail to meet the food requirements for all of the bears all of the time. Some of the reasons derive from how we manage the land. As forests are logged off, the second growth provides short-term food surpluses that lead to increased bear populations. But the overharvesting of hard-mast trees has a troublesome long-term effect: shortages in the forest's nut crop, a staple for bears. So, the bears' numbers increase as their staple mast crop grows less reliable.

Summer's vegetative foods would normally act as a buffer when the forest's nut crop fails, but those foods tend to grow in valleys, precisely where humans like to build their houses. With much of their critical summertime habitat destroyed or impaired, bears often have little or nothing to eat during food shortages or failures. And their search for what remains often brings them in close contact with residential properties. In New Hampshire, their preferred summertime food, jack-in-the-pulpit, draws bears quite close to homes. Once there, they can easily smell the food available in nearby yards, or follow other bears' scent trails to bird feeders or garbage cans.

In poor food years, when bears experience increased social aggression over diminishing wild resources, more and more bears will be willing to take the risk to enter backyards, farms, or areas with dumpsters to access food, even if they've experienced some prior diversion tactics. These can be dangerous years for bears.

My own decision to offer Squirty and other bears I've raised supplemental food was motivated by conditions like these. To keep them from being shot at a bird feeder or in a cornfield or lured by a bait, I decided to bring them food in the woods—far away from where they would associate it with homes or backyards. Once a bear gets food at a human residence, it gains knowledge about where the food comes from and about the behavior of the individuals supplying it. But a bear that finds human foods in the forest gains no knowledge about where the food came from or where else it could be found. So, remote diversionary food sites do not increase a bear's knowledge about how to get food from humans.

In that respect, feeding bears in the woods at carefully chosen locations in special circumstances can keep those bears out of trouble without giving them incentive to travel to places where they could encounter trouble.

The concept of supplemental feeding usually brings out an "over my dead body" response from wildlife managers. But the idea of mitigating bear nuisance behavior with food makes sense in some situations. Providing food buffers so that bears don't have to risk entering human zones in slim food years would not only decrease the cost of handling nuisance bears, it would also reduce the number of complaints, and the number of unnecessary bear deaths.

One recent spring when one of our collared wild females, LG2, emerged from her den with three small and undernourished yearlings, she spent three weeks at her den entrance conserving energy by not looking for food that she knew did not exist. When the green vegetation first became available, she moved down into the wetlands she had inhabited the previous year. She quickly discovered a large cache of high-quality food in the unattended garbage of a family that had gone on vacation. An expanded search of the neighborhood revealed a mix of wetland vegetation, bird feeders, and garbage. My first intervention was an attempt to persuade the members of that neighborhood to change their behavior by proactively removing birdfeeders and securing their garbage. I was somewhat successful, but that didn't solve the problem.

My next intervention was to use the human-dominance method. This worked very well in the yards from which I chased her. She never returned to them. However, she was still driven by the need to find food for her starving cubs and simply expanded her home range, finding bountiful supplies of birdseed and garbage. Every fifth or sixth home was feeding birds, deer, bear, turkey, or had unsecured garbage. She entered open garages to get garbage and sacks of birdseed. Her activities took place during daylight hours, and sightings placed her as much as five miles out of her normal home range.

I knew how to stop her, but my research license specifically says I'm not supposed to feed these bears. Finally, responding to calls about her was taking way too much of my time, so came my third intervention. I knew from previous experience that if I offered her food with less risk I could prevent her from bothering people. I entered the woods near where she was getting in trouble. I sprayed a large pine tree with citronella bug spray to attract her and put down a gallon bag of corn. She took it and I increased the amount to two gallons of corn a day at the cost of a dollar a day. My sister Phoebe took over the task of feeding her, which took only 10 minutes a day. My plan was to end it as soon as her mate arrived for the breeding season. It worked and the calls ended. After three weeks, we stopped giving her corn. Once the need to get good food to her cubs was over, she was content feeding on natural foods.

None of this would have happened, however, if human food wasn't outdoors to attract her. So, the ultimate solution remains in the hands of residents. Those who keep their yards food-free won't have problems.

Backyard Chickens, Bees, and Other Livestock

Since the downturn of the economy in 2008, there has been a resurgence of backyard farming. The number of people raising chickens for eggs and meat has risen, and this has created some new problems for both black bears and grizzlies throughout the United States and Canada. Sixteen of the twenty-nine orphaned cubs we received in 2012 had their mothers shot at unprotected chicken coops or beehives. Tragedies like this could be avoided by surrounding the coop or hive with an electric fence and routinely applying peanut butter, bacon grease, or another strong-smelling, sticky food to the wire with a sponge. You could also drape a piece of bacon over the wire or modify a sardine tin to hold a small amount of food and hang it from the wire. Bears check out all new smells with their tongues, and once zapped they have no more interest in what is on the other side of the fence.

Some states have passed laws making it illegal to shoot a bear in defense of property if no measures have been taken to protect livestock or bees. These proactive measures solve problems rather than allowing bad practices to continue.

Dealing with Bear Encounters

With bears and people encountering each other more and more frequently, it is essential to understand how to properly handle an unintended meeting in the backyard or on a hiking trail. The vast majority of all bear aggression toward people is protective, not predacious, and it is entirely possible for people to manage these protective encounters without injury. A key to doing so is to understand how bears communicate.

You can refer to chapter four for an overview of bear body language, sounds, and behaviors. But the most important thing to understand is that when a bear wants to intimidate you, keep you at a safe distance, or otherwise modify your behavior, it will square off its lips—drawing them forward so that they appear square and the face looks long. Then it will perform any of the following behaviors in varying degrees of intensity: chomping its teeth or lips, snorting or woofing (blowing air through the nose or mouth), huffing (inhaling and exhaling air rapidly), swatting, or false charging. These are actions that bears take to help reduce the chance of attack whenever two unfamiliar individuals come together. However, this behavior does not reflect the bear's true mood. Bears are able to turn this behavior on and off like a light switch. They are simply trying to delay confrontation long enough for communication to take place.

Moods, on the other hand, come and go very slowly. It is therefore necessary to analyze the bear's mood when it is not displaying these behaviors, its intentions when it is, and then to apply both to the context of the situation. This may be a tough concept to apply in the field, but a necessary and important one to understand. Being faced with a bear that false charges or bluffs is actually a good thing as it means you have time to analyze the bear's intentions and modify its displeasure or fear.

How do you know the bear is false-charging and not attacking? The false-charge is done in combination with other bluff displays, like chomping, huffing, and snorting. Depending upon the situation, this usually reflects the bear's desire to delay or avoid direct confrontation. For instance, when Squirty has cubs and wants me to leave, she may false-charge, letting me know her intent without having to attack. Wild bears have been known to use these displays to motivate people to drop food or knapsacks.

However, if you find yourself in such a situation and act in a reckless manner while the bear is within critical distance–as when a bear holds its ground and displays rather than flees–you can escalate this kind of situation into an attack. Reckless behavior would include breaking sticks, yelling or screaming, making yourself big by raising or waiving one's arms, or basically doing anything in which you could not anticipate a correct response. A safe response would be to de-escalate the situation by standing erect and speaking softly to the bear, thus signaling to the bear that you are dominant but not a threat.

When you have an encounter with a bear, it is always important to try to put yourself in the bear's shoes. Does the bear have any reason to harm you? Have you provoked the bear intentionally or unintentionally? Is the bear already nervous about other bears in the area? Remember that bears, like all other animals, including humans, have four major drives: hunger, love, fight, and flight. These drives are usually in conflict with each other.

It is also important to assess just how close you are to the bear. While it's always important not to take any action that leaves you unable to predict the reaction of the bear, it is particularly important if you and the bear are in close proximity (normally less than twenty-five feet) and the bear appears reluctant to leave. This twenty-five foot distance is known as the "critical distance" outside of which bears and many other animals are likely to flee. Within this distance, they are hesitant and uncertain as to whether they should act in self-defense or flee.

A conflicted bear in this situation will act as described above. My advice is to stand erect with eyes toward the bear. Do not attempt to stare the bear down but rather maintain a normal facial expression and speak softly. Standing erect and keeping eyes toward the bear will keep him or her honest. Bears, like dogs and humans, may choose to enforce dominance when the opportunity arises. If you show weakness (by lowering your eyes, turning your back to them, lying down on the ground, or showing fear), it increases the chance that they will take advantage and advance on you.

Think of someone you know who has been bitten by a dog. That person has probably been bitten more than once. Perhaps more commonly, your own dog becomes aggressive only toward certain people or dogs and not others. In both cases, the victim is sending submissive signals. Fear is your worst enemy when working with animals, because they can sense it. If you are fearful, you are also unpredictable, and being unpredictable makes you a threat to the animal before you. On the contrary, speaking softly (not the words, of course, but the tone) conveys a message of appeasement the bear will understand. Speaking softly will also help to calm you down. Bears are able to read our emotional communication and can determine whether we are a threat or not.

My advice to keep your eyes on the bear conflicts with almost every other message given about what to do when you are in close proximity to a bear. I look at the bear to remain dominant while I decrease the threat level with my voice. Others will argue that you should avert your stare because a direct stare is aggressive and may provoke an attack. My experience tells me that this is not the case with bears. Animals that live in group-social environments often have hard, top-down hierarchies. A stare at an alpha chimpanzee or wolf may be perceived as a challenge to its position of authority. Bears are different; they interact and cooperate with strangers on a regular basis and are used to negotiating with unfamiliar individuals.

The bear that gets too close is usually a sow with cubs. Her concern is the threat you present. She is perfectly capable of assessing that threat. Give her a chance, and she will walk away from you, sometimes even leaving her cubs up a tree nearby. I have been inside that critical distance with more than thirty wild sows with cubs, been false-charged and circled (bears circle to check scent, to see who you are), and have then gone on to peacefully spend up to two and a half hours with them. Every female will exhibit a different level of aggressiveness. Most of the wild sows and cubs I have encountered ran, hid nearby, and waited for me to leave. There are many myths about sows with cubs–the prevailing one being that if you get between a sow and its cub you are toast. The reality is that sows with cubs have been responsible for only 3 percent of the fatal attacks on humans in the last 109 years. Their cubs are usually safely up a tree when close encounters occur. Having preconceived ideas in your head at these times will only make it more difficult to control the situation.

So, imagine that you meet a sow with cubs on a trail. You are torn between running and standing your ground. She is torn between running and defending her cubs. She would like to run, but her cubs are up a tree. She chooses to display aggressively in an effort to prevent you from attacking. You would like to run, but you know that she can run faster. You try to relax, knowing that fearful behavior could be seen as a threat. You speak softly to her as a gesture of appeasement. She acknowledges your gesture by reducing the intensity of her displays. Be patient. Eventually, she will stop displaying altogether and her true mood will be revealed with a relaxed facial expression. Slowly, she will walk off. For obvious reasons, the drive to escape is generally stronger than the drive to fight. She knows a fight could leave her wounded or dead.

Yelling and screaming to drive away a female bear away, on the other hand, may inadvertently frighten her cubs and escalate the situation. Squirty, who is fairly aggressive and treats me like another bear because I raised her, has taught me a few things about being around sows and cubs. She is much more aggressive when her cubs are small than when they are older. The one thing I don't want to do is scare her cubs. If I do, I get an immediate aggressive response.

If you meet a male bear, the situation may go somewhat differently, but the same advice about handling the encounter applies. Male bears are much more likely to run off than be caught inside the critical distance. If you see a bear coming in your direction, it is a good idea to let it know that you are there. Bears read scent in the wind, but sometimes the wind is coming from the wrong direction and a bear may be completely unaware of your presence. Let the bear know that you are there by moving, talking to it, or making other noise; it will run off.

But there are situations where attacks are more likely. A bear that is surprised while eating–or while its senses are otherwise compromised–may strike out without warning. For instance, a bear feeding on a carcass is highly concerned that other bears may be attracted to the carcass by smell and is preconditioned to attack. A person who suddenly appears in this situation may trigger that preconditioned attack.

Predacious attacks also have a different nature. They are extremely rare, averaging fewer than two a year over the past 109 years, but they account for roughly 85 percent of all the fatal black bear attacks and are generally carried out by males. Forty-nine of the fatal attacks occurred in Canada and Alaska and only fourteen in the lower forty-eight states, despite a higher population of both bears and humans there. Predacious attacks tend to come from behind; they also tend to be silent to avoid detection. But being alert in the woods may help. In India's Sundarbans region, between 50 and 250 people a year are lost to tiger predation. The victims are mostly woodcutters who are so involved cutting wood they become easy prey for the tigers. The wardens, who also work in the forest, are rarely taken because they stand erect and watchful.

Bears are highly tolerant of humans. There are hundreds, perhaps thousands, of bear encounters every year where humans do everything wrong without any negative response from the bear. It's important to remember, though, that in the vast majority of cases, black bears are dangerous only if you make them so. The situation is in your control; they tend to signal their intentions, and you can modify your own behavior to influence theirs.

INDEX

NOTE: *CI* pages refer to color insert.

Index

Index

ABOUT THE AUTHOR

Benjamin Kilham has been researching and living with black bears for nearly twenty years. He has become an expert in black bear behavior, as well as in rehabilitating orphaned and injured bears and reintroducing them to the wild. He is invited to lecture all over the United States and internationally. His previous book is *Among the Bears: Raising Orphan Cubs in the Wild.* Kilham and his work with black bears have been featured in five internationally televised documentaries, including National Geographic and Discovery Channel features, and he has appeared on *The Today Show*, *Good Morning America*, *ABC Nightly News*, *ABC Nightly News International*, *The O'Reilly Factor*, *Fox News Daytime Edition*, *Inside Edition*, *The David Letterman Show*, *NBC Nightline*, *CBS Coast to Coast*, *Canadian Broadcasting Company Nightly News*, and various other shows, as well as National Public Radio and a host of other nationally broadcast radio shows. He lives in Lyme, New Hampshire.

About the Foreword Author

Temple Grandin is one of the world's most accomplished and well-known adults with autism. She is a professor of animal science at Colorado State University and has long been a leader in developing humane livestock-handling techniques. She is the author of six books, including the national bestsellers *Thinking in Pictures* and *Animals in Translation*. A past member of the board of directors of the Autism Society of America, Grandin lectures to parents and teachers throughout the United States on her experiences with autism. Her work has been covered by numerous news outlets, including the *New York Times*, *People*, National Public Radio, and *20/20*. She was named one of *Time* magazine's one hundred most influential people of the year. The HBO movie based on her life received seven Emmy Awards.